Revitalizing U.S. Electronics

Lessons From Japan

John L. Sprague

Butterworth-Heinemann

Boston London Oxford Singapore Sydney Toronto Wellington

Recognizing the importance of preserving what has been written, it is the policy of Butterworth-Heinemann to have the books it publishes printed on acid-free paper, and we exert our best efforts to that end.

Library of Congress Cataloging-in-Publication Data

Sprague, John L., 1930–
Revitalizing U.S. electronics : lessons from Japan / John L. Sprague
p. cm.
Includes index.
ISBN 0–7506–9223–5 :
1. Sprague Electric Company—Management. 2. Electronic Industries—United States—Management—Case studies. 3. Electronic industries—Japan—Management—Case studies. I. Title
HD9696.A3U614 1992 92–54696
621.381'068—dc20 CIP

British Library Cataloguing-in-Publication Data
A Catalogue record for this book is available from the British Library

Butterworth-Heinemann
80 Montvale Avenue
Stoneham, MA 02180

10 9 8 7 6 5 4 3 2 1

Printed in the United States of America

Table of Contents

Dedication

I dedicate this book to my father, Robert C. ("R.C.") Sprague, who died at the age of 91 on September 27, 1991. Educated at the U.S. Naval Academy, the Naval Post-Graduate School, and at MIT, "R.C." founded what was to become the Sprague Electric Company in the kitchen of his home in Quincy, Massachusetts in 1926 while on active duty in the U.S. Navy. The first product was a radio tone control which he had invented. However, it was an entirely new miniature fixed paper condenser, or capacitor, that was part of the tone control that created the future of his company. Sprague Electric grew to become one of the world leaders in electronic components due to his genius, the outstanding executives he was able to bring into the company, and continuing leadership in materials technology.

Active throughout his life in numerous endeavors both within and outside of the electronics industry, he also served his country as an expert on continental defense during the Eisenhower years. Of his many honors, probably the most prestigious was his 1978 receipt of the Electronic Industry Association's highest award, the Medal of Honor, for the second time. No one else has ever been so honored.

Having helped build an industry, he served as a superb role model for legions of executives who knew him and especially those who were lucky enough to work for him. His passing leaves a void that will be impossible to fill.

Acknowledgments

So many people helped me out in the writing of this book that it is difficult to give full credit to all of them. The "Notes and Sources" section at the back of this book provides a list of some of the most influential. Of this group, I would especially like to thank Dr. Sei-ichi Denda of Konica for his insightful inputs over a number of years; Darnall Burks and Mikiya Ono of Mitsubishi Materials along with Galeb Maher of MRA Labs for much of my education in electronic ceramics; Jack Driscoll of Murata Erie North America for helping organize my highly fruitful visit to Japan in June of 1991; and Peter Maden of Sprague Technologies for his unflagging defense of the U.S. passive components industry. I would also like to thank Dick Morrison and his staff at Allegro Microsystems, formerly the Sprague Semiconductor Group, for their inputs on the semiconductor history of Sprague Electric, and Lennox Lee of Sprague Technologies for his highly important industry marketing data. Additional gratitude goes to Ed Getchell, president of Venture Founders, for his insight on the venture business and advice to find a corporate partner for NEWCO rather than equity.

Last, but certainly not least, I must thank my family, especially my wife, Jid, who over two years read parts of the manuscript and, without complaint, accepted a hiatus on much of our social life; my daughter, Cathy, who taught me handbuilding pottery; and my son, Bill, at Smith Barney who shared his thoughts on the U.S. economy and cost of capital. I am also indebted to my cousin Peter Sprague, chairman of National Semiconductor, for numerous lively debates concerning the U.S. semiconductor industry. While we didn't agree on everything, we certainly stimulated each other's thinking.

Introduction

When I was first approached about writing a book comparing United States and Japanese competitiveness in the electronics industry, I questioned what new I could add to the volumes that have already been written on the subject. After all, I certainly am not a professional writer, nor am I associated with a large and prestigious consulting firm. I am not a learned academic or dean of a well known business school.

On the other hand, I do have some reasonable credentials, and many of them are of a practical nature directly applicable to the subject. For one thing, I have some pretty good genes relative to the electrical and electronics fields. For example, my grandfather, Frank J. Sprague, was a well recognized inventor and is considered the "father of electric traction." His son—and my father—"R. C." Sprague started the Sprague Specialties Company (later to become Sprague Electric) in the mid-1920s by inventing a radio tone control in his kitchen. His later development of a radically new capacitor to make the control work led to Sprague Electric's position as the world leader in capacitors by the 1960s.

My education led me down much the same path my father had taken. I received my A.B. degree in chemistry from Princeton and followed that with three years of active duty in the U.S. Navy during the Korean War. Two years of my service was spent in shipboard electronics. After the Navy, I received a Ph. D. from Stanford in physical-inorganic chemistry. My doctoral thesis was on semiconductor materials and devices.

After leaving graduate school, I joined Sprague Electric as a research scientist. I spent nine years in corporate research and development, eight years as head of the semiconductor division, five years as president and chief operating officer, and finally six years as chief executive officer. Between 1977 and 1987, I traveled extensively in Asia, partly as a result of serving for ten years as a director of Nichicon-Sprague (N-S), a joint venture between Sprague Electric and the Japanese firm Nichicon.

During my 28 years with Sprague Electric, I learned how to "partner" as well as compete with Japanese companies in world markets. As I thought about such experience and my background, my writing a book about how U.S. and Japanese corporations compete in electronics didn't seem so far-fetched after all—but only if I could support my conclusions with detailed personal experience and make the narrative interesting. Thus, once I decided to proceed and my outline was accepted by my editor, this book began to develop a much more personal tone than is generally found in such treatises.

I wanted to discuss the entire commercialization process, from innovation to manufacturing to serving the market place. I also wanted to deal with some of the misconceptions we as a nation have concerning the relative strengths and weaknesses of the United States and Japan. For example, it is true that the United States remains the world's most innovative nation. All too often, however, Japanese firms are the ones that capitalize on such innovations. But it is frequently U.S. corporations themselves, through such methods as technology licensing and partnering arrangements, that have made it possible for Japanese corporations to do so. And Japanese skill in commercialization of technology and manufacturing is not the result of some government-directed conspiracy but instead is a matter of emphasis and resource allocation, starting with the Japanese primary school system.

I am a "free trader" by philosophy. However, as my research progressed I was surprised at how unequal the playing field really is, especially when it comes to the relative ease Japanese firms have in penetrating the United States market compared to the problems U.S. firms have in serving Japan. I also agonized over whether or not it is really possible, or even appropriate, for the U.S. government to develop an effective industrial policy. I was also surprised—and somewhat relieved—to find cracks developing in the Japanese monolith. All these issues plus many others are discussed as the book progresses.

However, I am getting ahead of myself. My education concerning Japan really began in detail when I made my first business trip there in 1977. The description of that trip that follows clearly shows how much there is to learn about how a different society and economic system operates, lessons best learned through personal experience.

John L. Sprague

CHAPTER 1

Sprague in Japan: Expectations, Realities, and Some Lessons

Viewed from the airport lounge window, the airplane at the end of the jetway looked like a reject from the Boeing production line. Fore and aft, it was a conventional 747. Much of the center of the aircraft was missing, however, giving it a short and squat appearance. It was a 747SP (special purpose) jumbo jet, and it was the airplane that was going to take me on my first trip to Japan. At the time, the 747SP was the only commercial jet that could fly non-stop from the United States mainland to Japan. This added performance was accomplished by removing approximately 100 seats and part of the central section from a conventional 747. I flew the 747SP many times over the next several years. Each time I silently applauded the American aerospace industry for the fact that such an apparently unbalanced aircraft could fly so well.

It was early 1977 when I made my first trip to Japan. At that time, I had recently been made president and chief operating officer of the Sprague Electric Company, a major worldwide manufacturer of electronic components. Sprague had a rich heritage in capacitors and was also a niche supplier of semiconductor integrated circuits (ICs). I had managed the IC business unit for eight years before I became President. The primary reason for this first visit to Japan was to meet with our Japanese semiconductor sales agent, Sanken Electric Company. Unfortunately, Sprague was not then enjoying the same success as Boeing when it came to successfully moving its products from the United States to Japan.

Because of the length of the flight, I was armed with voluminous reading material, including a detailed status report on Sanken, internal Sprague reports, several technical magazines, and two books of a lighter nature. In addition I had brought an "instant Japanese" course, hoping to learn at least a few words. This would turn out to be of little use, except to provide me with a few polite phrases. As with most Americans, my lack of knowledge of the Japanese language was to prove a severe disadvantage, especially during business negotiations. Most of the Japanese I have met understand at least some English and a surprising number are fluent. Some have studied or worked in the United States, or learned the language through self study. In addition, English is required in the Japanese school system. (This was brought home to me while visiting a shrine in Kyoto on a later trip. A young student broke away from her group, came up to me and proudly said, "herro, how are you??" Pronunciation of the letter "L" is a problem for many Japanese, since their language has no fully equivalent sound. After I replied, "I am fine, how are you?", she ran giggling back to her friends and the little girls in their Navy-like uniforms moved off en masse down the street.) However, being able to speak a language is not the same thing as being able to understand a culture or society. Most Japanese I knew still had nearly as much difficulty understanding our society as we did theirs unless they had actually lived in the United States for some time.

Another part of the "baggage" I carried was a mixed set of preconceptions about Japan, some negative and some positive. On the negative side there was a certain amount of distrust relative to Japanese business practices, primarily a subliminal carry-over from the attack on Pearl Harbor and World War II. Even over 50 years later, the anniversary of the "day of infamy" resurfaces these feelings like a running sore that will not heal. On the other hand, Japanese executives that I had met in the United States prior to this trip were—almost without exception—intelligent, articulate, good businessmen, and a pleasure to be with. Because of these conflicting notions, I was not sure what Japan would be like. I anticipated an extremely crowded country populated by industrious people with an extraordinary national pride. Those individuals I knew who had actually lived in the United States admitted that life here was much pleasanter than in Japan. Still, they all wanted to eventually return home. Their primary concern was the westernization of their family, especially the children.

Since I sleep poorly while flying—despite wine, food, movies, reading, and conversation with my neighbor—the flight seemed endless. I departed Kennedy airport in New York late on a Saturday morning. My plane flew "with the sun" and, because of the international dateline, arrived late Sunday afternoon, some 14 hours later. Even before landing in Tokyo, I had learned my first lesson about doing business in Japan: try to arrive on a Saturday, thus allowing a day of recovery and relaxation before getting to work.

I arrived in a stupor, not at all sure how I could manage my meetings the next day. Customs was short, since I had carried everything on board. The trip into Tokyo was eased by the fact that Kerry

Enright, Sprague's sales manager in Japan, had met my plane. However, we got caught up in the evening weekend traffic back to the city and it was slow—stop and go—the entire way. I later learned that traffic around Tokyo is like this most of the time. The Tokyo drivers were apparently used to it. They flashed their lights a great deal, but never used their horns.

After what seemed like hours, we arrived at the Tokyo Hilton. Unlike most of its American counterparts, it was a surprisingly beautiful hotel. Nearby was the Imperial Palace. I shared a light dinner with Kerry and agreed to meet him the next morning for breakfast. After that, we would head for the Sprague office and prepare for the meeting with Sanken. I returned to my room, which was quiet, pleasant, and whose atmosphere was accented by the oriental decor. I headed straight to bed and fell instantly to sleep, only to wake several hours later and sleep fitfully until morning. I learned my second lesson: a sleeping pill is a big help the first night in Japan.

At breakfast, I was slightly dizzy from lack of sleep. (I learned a third lesson: Japanese ham and eggs are best left to other diners!) Afterwards, Kerry and I took a taxi to the office. Our taxi was cleaner than any American equivalent I had ever experienced, and was in fact cleaner than most private cars. It was a good place to take my first look at Tokyo in the daytime.

What I saw was a huge, clean, sprawling city of contrasts, very crowded, and filled with hurrying people and automobiles. Modern high-rises, old houses, pagodas, small shops, little parks, and giant stores all vied for the much too limited space. The Sprague office was down a small side street and, much like Tokyo itself, spartan, congested, and a beehive of activity. Compared to similar offices in the United States, it was more like what you would expect of a new, cash poor start-up than the Japan headquarters of a major international corporation.

The cramped office didn't seem to bother the six or so employees. They were all young, bright, energetic, smiling and very happy to meet "the boss." As we got to know each other, they would also prove to be surprisingly outspoken on what was good about Sprague Electric—and what was not. Their criticisms focused on how their customers were being hurt by missed delivery promises in addition to a rash of quality problems related primarily to cosmetic variations in visual specifications. I was not surprised about the delivery complaints. On my regular visits to sales offices around the United States, missed deliveries with most Sprague Electric product families, including capacitors, were the source of repeated complaints. However, quality problems were almost non-existent. You might have difficulty in getting a Sprague part, but once you did it was invariably of higher performance than the competition and almost never failed. I could not understand why our customers in Japan would have complaints about quality while our American customers did not.

In preparation for the afternoon meeting, Kerry and I reviewed the history of the Sanken relationship. After several years of attempting direct sales of ICs to the Japanese market with poor results, Sprague had decided to try working through a trading partner or sales agent. At the time, Sprague's semiconductor division was just too small to even consider going directly into manufacturing in Japan. However, we felt that somewhere down the line a joint venture along the lines of Nichicon-Sprague (N-S) might be in order. N-S was Sprague Electric's highly successful Japanese venture in solid tantalum capacitors with Kyoto-based Nichicon Capacitor. As a matter of fact, Sanken's name came to us from Nichicon, who we had first approached but who had shown no interest in the semiconductor business. With semiconductor revenues under $100,000,000, Sanken seemed an ideal partner. They were already manufacturers of discrete power transistors and hybrid circuits, and were also representing Fairchild Semiconductor in ICs unrelated to the Sprague products. In addition, they seemed anxious to sell the Sprague line. On the other side, the relationship would allow Sanken to serve their customers with a new family of state-of-the-art semiconductor components.

The initial products involved were Sprague's proprietary series ULN 2000 Darlington arrays, a family of high voltage, high current arrays. They were comprised of seven silicon NPN Darlington pairs integrated into a common monolithic silicon substrate. The Darlington configuration (see Figure 1-1) behaves like a single transistor with very high gain. The circuits were designed to provide the current carrying and voltage-sustaining capability required when logic circuits must control peripherals such as relays, solenoids, card punches, printer heads, tungsten filament lamps, and LED displays.

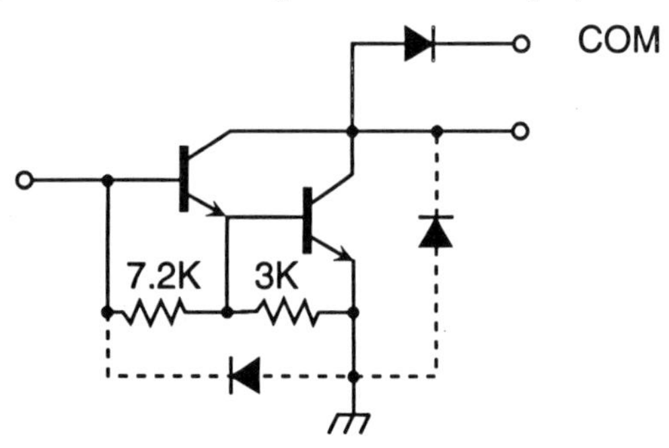

Figure 1-1: Typical Darlington configurations in Sprague Electric (now Allegro Microsystems) ULN 2000 series.

At the time, there were five different circuit configurations (identified as 2001 to 2005) designed to interface with different logic systems, including DTL, TTL, PMOS, and CMOS. Although deceptively simple at first glance, integrating such devices on a single IC chip had never been tried before because of the high current and voltage ratings involved. In fact, the ULN 2000 devices had grown out of a previous Sprague family of chip-and-wire hybrid integrated circuits. Realization required new diffusion profiles and packaging schemes, and for several years Sprague dominated the worldwide market. Computer peripherals represented our major market opportunity. American electronic equipment manufacturers and Japanese customers such as Seiko, Toshiba, Hitachi, NEC, Sharp, and TDK saw real advantages in using them to replace the many discrete devices serving the same interface functions in conventional board layouts. This let our customers reduce printed circuit board "real estate," increase reliability, and reduce cost. Today, this family is a true commodity, with worldwide unit sales approaching 100,000,000 annually at average prices under $0.25, compared to four to five times that amount when they first hit the market. Figure 1-2 shows a photo of the ULN 2005.

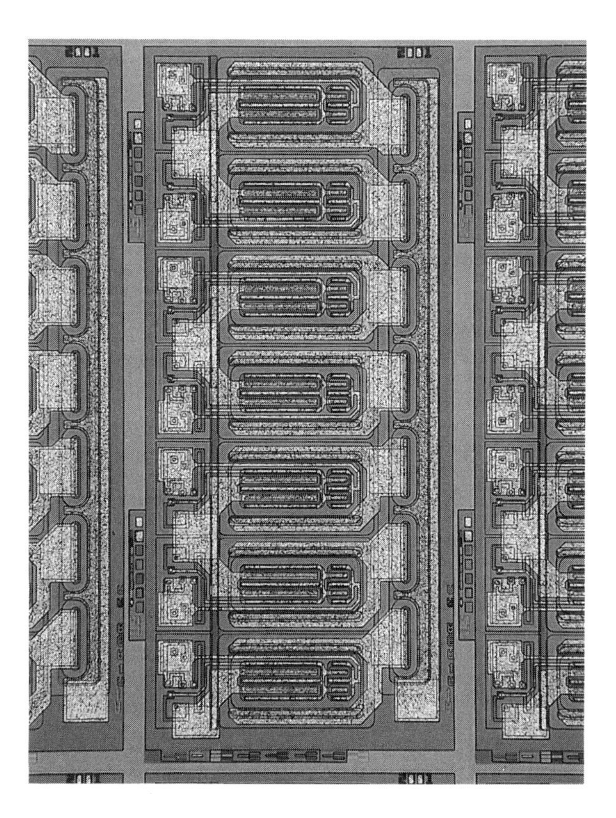

Figure 1-2: Photo of the ULN 2005 chip.

Ironically, Sprague's dominant position and customer acceptance of the product family were the major causes of our early problems with Sanken.

At the time of my first visit to Japan, the Sprague–Sanken relationship was in big trouble. Because of demand, Sprague was unable to meet its commitments anywhere, including the United States. But there was no question that we tended to serve our American customers first. Partly this was due to geography and the fact that most American customers also purchased other Sprague component families. However, it was also partly due to a lower overall priority Sprague gave the Asian market relative to both the United States and Europe. We had taken some steps to deal with the problem, such as assignment of a person full time to interface with Sanken at our U.S. factory in Worcester, Massachusetts. Nightly expedite calls from Japan had also helped some. Despite this, we were literally months behind our commitments to Japanese customers. And there were the previously-mentioned quality problems, primarily involving mechanical and visual defects. If there was any consolation for us, it was that Fairchild was apparently in even more difficulty. As a result, Sanken and their customers had developed a very poor initial impression of American semiconductor suppliers.

Naively, I really expected to be able to handle these problems with Sanken much as I did with our U.S. customers. At our morning review, Kerry and I decided to take the direct approach and to explain what steps Sprague was implementing to increase production and improve quality, to suggest the placement of a buffer inventory in Japan (if we could ever build it up), and to be as responsive as possible to their complaints. Because the product line was so new, proprietary, and provided so much added value, we were almost always able to juggle deliveries and to work our way out of each specific customer problem somehow. Experience had shown that being honest, a good listener, and doing the best you could usually bought sufficient time to solve most major difficulties. I was really unconcerned about the quality issue since our American customers had experienced little or no apparent problems with exactly the same devices. Kerry also seemed untroubled. (In hindsight, it was probably because it was I and not he who was about to take some heat.) Thus, our mood was relatively upbeat as we drove to the Sanken factory on the outskirts of Tokyo. I always approach customer meetings with some apprehension, so in that respect this one was no different (my mood was not helped by some minor nausea caused by jet lag). Nevertheless, I was counting on the well-publicized Japanese politeness to ease me through my first business meeting in Japan.

I was about to get my biggest lesson of the trip.

After signing the register in the modest and unimpressive lobby, we were ushered into a small, dark waiting room and served tea by two uniformed young women. We never saw their faces because they moved about the room in a continuously bent position. We sat there for more than half an hour, and our only companions were the faces of grim-faced men hanging on the walls. It did not seem an auspicious beginning.

Finally, we were taken to a much larger room filled with a square conference table. Around it sat fifteen unsmiling men. They were the board of directors of Sanken Electric. When Kerry and I sat down in the only two empty chairs, all faces turned toward us with what could only be described as expressions of contempt.

As soon as we were seated, a slender man on my right—I believe it was the then President, Mr. Fukuhara—stood up and slammed his fist down on the table. He shook everything on the table and my composure. He started speaking rapidly in Japanese and continued for more than fifteen minutes, his voice rising and falling with passion. As if in unison, the other members of the board nodded in agreement. While I did not understand a word he said, I certainly understood the tone. Finally he stopped, glared at me, and nodded to a young man next to him (who, I noticed, had started to perspire profusely). Acting as the interpreter, and without looking at us, he summed up the tirade as follows: "Honorable president says we have a problem!"

Fukuhara continued in a similar manner for more than an hour, stopping for brief translations like "Sprague Electric is not meeting its commitments," "our customers are very unhappy with the quality of the product they are receiving," "there have been line shutdowns with our customers because of late deliveries," and "Sprague does not understand what is required to serve the Japanese market." This one-way dialog seemed to go on endlessly until he suddenly sat down. I realized it was now time for my reply.

Using Kerry as the interpreter, and speaking very slowly, I began to describe the evolution of our product line, explaining it was very new technology and that we were still having yield problems. I also outlined all the steps being taken to add capacity and improve deliveries. I pointed out that as far as quality was concerned, our American customers seemed perfectly satisfied. As I talked and looked around the room, only one person seemed to respond to what I was saying. That was Dr. Sei-ichi Denda, who had considerable experience in the IC business and who was directing this effort at Sanken. While he occasionally nodded in agreement with my comments, he was the only one in the room who did. To everyone else, Sprague had made commitments which were not being met and, as a result, Sanken customers were being hurt. From our point of view, however, we were doing the very best we could. Like many U.S. suppliers, we were operating on the philosophy that the best way to get a business going was to get an order, ready or not. As a niche semiconductor supplier driven by the desire to be first to market with technically superior products, the trade-off between early product leadership and being completely ready presented a real dilemma throughout my involvement in the business. (A recent letter from Dr. Denda, now at Konica, gives his view on the subject: "Perhaps the reasons for the small share U.S. semiconductor manufacturers have of the Japanese market is due to continuing delivery problems and inferior quality compared to Japanese competition." I guess this shows some things never change.)

Our dialog continued for over three and one half hours until there was nothing left to say. I sat down, exhausted. I promised to look into all the problems personally and guaranteed to do better.

The ride back to the hotel was a somber and mostly silent ride. Furious with Kerry, I asked how he could have let me go into such a confrontation without the slightest indication of what was ahead. He sheepishly replied that our meeting was unlike anything he had ever experienced in Japan and, while he knew there were real difficulties, he had never expected anything like what had happened. We finally concluded that at least some of Fukuhara's tirade was for effect, both on us and to show his directors and managers what a tough man he was. At least as far as Sprague was concerned, it worked. I had made commitments and, regardless of other priorities, they would be met. (I have always wondered why more salespeople in trouble with a factory or division don't try to get the president or CEO involved. As the best peddlers have all learned, there is no quicker way to get problems solved, generally much to the chagrin of the offending business unit.)

Thus ended my first business meeting in Japan. It was a meeting that would set the stage for all our future dealings there.

When I returned to the United States, we worked closely with Sanken and began a program that was to lead to tremendous improvement over the next several years. It did not happen at once, but instead one step at a time as we collectively learned what was required. First, as yields improved and the market stabilized, we were able to begin to get an overall handle on delivery requirements. Japanese customers were given equal priority with everyone else, and Sanken was able to gradually build a three month buffer inventory. Since the requirements in Japan on quality and detailed response to quality problems were more stringent than in the U.S., over the 1978-1979 period Sanken created an in-house final test and burn-in facility as well as the capability to do detailed failure analysis. In effect, they began to "Japanize" the American manufactured products to meet the requirements of the Japanese market. In the process, they were also able to gradually increase their own capability in back-end manufacturing of ICs. As a result, I began to have an uneasy feeling that—although the program was going well and our sales in Japan increasing—somehow we were no longer in control.

Feedback from Kerry and subsequent visits to Sanken by other Sprague executives indicated that, as a next step, they were planning to request a technology license in wafer fabrication, probably during a trip I planned to make to Japan the following year. In anticipation, we held a number of sometimes acrimonious meetings at Sprague headquarters in Worcester to discuss the pros and cons of such a license. There was no question that it would provide immediate additional income and several members of our management team wanted to proceed. They were also afraid that if we refused, Sanken only had to turn to Fairchild with the same request to get what they wanted. I did not agree with this. I believed Sanken was after the power technology which, of the two U.S. partners, was only available from Sprague. (As far as I know, such a technology licensing request was never made to Fairchild.) I was also concerned about two additional points. First, transfer of such technology would be complex and time consuming and would interfere with all we had to do with running our own business. More importantly, I felt that all we would be doing in the long term would be to create another strong competitor. There were already several other aggressive Japanese companies supplying the line. They had obtained the technology through the complex and expensive process of reverse engineering devices obtained from Sprague-Sanken customers in Japan. The joint venture alternative was never really pursued, since we understood that this did not interest Sanken. Therefore, after lengthy debate we agreed that, if the request was made by Sanken, we would politely refuse to license our technology. I also decided that, no matter when or how the request was received, I would be the bearer of this decision.

My reception at Sanken at the beginning of my 1980 visit was very different from my first one. To a large degree, the original problems had been solved by our combined efforts. Along with other semiconductor products such as Hall effect sensors, Sprague was now selling close to $10,000,000 in semiconductors through Sanken into the Japanese market. Japanese competitors with low prices were now our greatest problem. As we shared tea together, President Fukuhara smiled and acknowledged how successful the program had become. He stated that Sanken would now like to take the next step—to license from Sprague the technology for front-end or wafer manufacturing. We discussed why they wished to do so, and what he felt were the advantages to Sanken compared to the current arrangement which I had said Sprague would be happy to continue. I am unclear about exactly what his stated reasons were, but I believe they related to giving them an advantage in new product development for their local customers. Finally, I politely stated that Sprague was unwilling to do this. I explained that once Sanken had this final capability they would no longer need Sprague Electric as a partner.

I was about to receive another lesson about doing business in Japan.

Fukuhara's face flushed with anger. In a rare moment of frankness, he said, "Why do you think we entered into this relationship in the first place?" The meeting summarily ended. Our association with Sanken did the same soon afterward.

Over the next several months Sanken moved quickly. First—and without our knowledge—they informed their customers that they were ceasing to represent Sprague because we were an unreliable supplier. Since Fairchild had already terminated their affiliation, Sanken was in effect exiting the IC business. In the meantime, as Sprague continued to fill existing orders, Sanken's customers began to shift to other, primarily Japanese, suppliers. Upon finally learning of Sanken's actions and intent, we attempted to retrieve the relationship but to no avail. Sprague then formed Sprague Japan KK (SJKK), and again attempted to go it alone. Since SJKK was only a sales arm, the results were very similar to what we had experienced before Sanken. The damage had already been done and the customer base lost. To this day, Sprague enjoys only a fraction of the Japanese semiconductor business they had at the peak of the Sanken program.

After our refusal, were Sanken's actions typical of Japanese business practices? Was my original feeling of mistrust well-founded? While I can only speak from personal experience, I think the answer to both questions is no. In a number of subsequent business dealings in Japan, I never again ran into anything similar. While the negotiations could be very difficult, once a deal was made it was always implemented openly and as agreed upon. So why did Sanken take the course they did? I believe that, because of their difficulties with both Fairchild and Sprague, they really felt they could not depend on any U.S. partner long term. Could the relationship have been saved in some form? Was a joint venture between Sprague and Sanken really impossible? Even with the benefit of hindsight, it is hard to answer these questions. I do feel that—even after the success of the original program—there was just too much mutual mistrust for the two companies to really act as equal partners. I think that Sanken believed that eventually they needed to be in charge of the entire business enterprise.

What happened to Sanken's goal of becoming a supplier of ICs? Despite the knowledge they already had gained and their own internal capability in discrete power transistors, they apparently were unwilling to take the long and expensive risk of trying to reverse engineer the Sprague IC process. But fate had an interesting twist in store. In a September 27, 1990 press release, Sprague Electric (now Sprague Technologies, Inc. or STI) stated the following: "Sprague Technologies, Inc. announced today that it had signed a letter of intent to sell its Semiconductor Group to Sanken Electric Co., Ltd. for $58,000,000 in cash. The transaction is subject to negotiation of a definitive agreement, certain governmental filings and approval by the boards of directors of both parties." The purchase and sale was closed on December 20, 1990. Since termination of the previous alliance a decade earlier, there had been virtually no interface between the two companies and, to the best of my knowledge, Sanken had not manufactured or sold any ICs in the open marketplace.

Although Sanken clearly wanted the Sprague technology, I doubt that acquisition of the Sprague Semiconductor Group was Sanken's plan from the beginning. Nor do I believe that they had been sitting in the wings for ten years waiting for Sprague's decision to sell the business. (Sanken's name was suggested by STI to their investment banker, Goldman Sachs, who was ultimately able to facilitate the transaction.) Still, Sanken's experience with Sprague must have helped with its deliberations and, despite the many problems associated with the original relationship, they must have thought the acquisition a good buy. With the transaction, Sanken now owns and controls the Sprague Semiconductor Group, which has been renamed Allegro Microsystems. This is very different from a partnership relationship with a U.S. company. While only time will tell, I understand the relationship is going well so far.

The period of Sprague's involvement with Sanken and its joint venture, Nichicon-Sprague, gave me a crash course in doing business in Japan and Japanese perspectives on the electronics industry. From 1977 through 1987, I visited the Orient more than two dozen times in different roles. Sometimes it was as a director of Nichicon-Sprague. At other times, I visited the Sprague operations in Manila, Hong Kong, Taiwan, and elsewhere. And there were other purposes—to visit Asian customers, to follow-up on our successes and failures in Japan, or to try and better understand how the Japanese operate in the electronics industry. Most of my visits were just as "educational" (although thankfully less bruising!) as my first visit to Japan and the initial meeting with Sanken.

What did I learn? For one thing, to the Japanese, a commitment is a commitment to be met, regardless of how trivial it might seem. The emphasis Sanken's customers placed on visual defects of Sprague ICs is a good example—to them, visual and cosmetic flaws represented poor workmanship and manufacturing practices. This concept is one many American companies have difficulty grasping. Perhaps this is why many (and probably most) Japanese companies believe local suppliers are more reliable than foreign sources and that they supply higher quality products. If a product is new and unique, Japanese customers will buy it from foreign companies. However, they would prefer to buy it from Japanese suppliers. Therefore, foreign companies must expect strong local competition to appear, especially if the product has high growth potential.

I also learned that using a Japanese partner may be the best way to initially penetrate that market. However, to the degree it is possible, it is important to try to clearly understand what the partner's long range goals are. As demonstrated with Sprague's experience with Sanken, be prepared for surprises. Based on the Sprague Electric experiences with Nichicon-Sprague and Sanken, joint ventures in Japan are far superior to straight technology licensing. If U.S. firms do decide to "go it alone" in Japan, such firms must understand that the Japanese have great staying power. Success in the marketplace by U.S. suppliers requires making a long term commitment. Many years may pass before financial success in our terms is realized.

I learned that Japanese corporations are fiercely competitive, especially among themselves. The image of a monolithic "Japan, Inc." is only partially accurate. And while politeness is cultural in Japan, do not expect it to have any impact on business decisions. As demonstrated in this chapter, it may be largely absent under certain conditions.

I discovered most Japanese CEOs are very strong leaders and, in my experience, dominant in their companies. This conflicts with the popular belief in the United States that consensus is the driving force in decision-making by Japanese companies. The primary role of consensus appears to be to guarantee agreement with the CEO, to spread the risk by assuring that any failure is shared by all, and as a polite way of saying "no." All too often I heard "we could not reach consensus" as an excuse.

In the chapters that follow, I will be expanding upon the ideas above, sharing additional conclusions and insights, and exploring what the American electronics industry can learn from the Japanese. This book will be both descriptive and prescriptive; I will describe what the current situation of the U.S. electronics industry is and how it got that way so I can explore ways it can be—as this book's title says—revitalized and made more competitive with the Japanese. I don't have the hubris to claim that this book is the definitive answer to the problems of the U.S. electronics industry, but I feel that my experiences both with the Japanese and in the United States electronic components industry (now totaling over three decades) equips me to offer advice others will find useful.

My primary vehicle for exploring the Japanese and American electronics industries will be the multilayer ceramic capacitor (MLC). One strong reason for this choice is because that is the area of my greatest experience and professional expertise. Another is that capacitors are even more ubiquitous than integrated circuits in electronics systems and devices. Capacitors are found in everything from portable radios to notebook computers to communications satellites, and even in systems built from discrete components (including vacuum tubes) instead of ICs. Perhaps most importantly, the process of development, manufacturing, and marketing of MLCs neatly encompasses many of the factors responsible for success—and failure—in the international electronics marketplace.

To understand where the Japanese and American electronics industries are today, we need to take a look at the scientific history of electronics. As it turns out, it is a surprisingly long history that challenges some widely-held assumptions about what it takes to succeed in electronics. That's the subject of the next chapter.

CHAPTER 2

The History of the Electronics Industry: Myths vs. Facts

As a nation, we in the United States have developed (almost by osmosis, it seems) an overwhelming impression that the U.S. leads the world in innovation. We also believe the Japanese created their post-World War II economy primarily on a foundation of basic science and technology developed elsewhere, especially in America. After all, didn't important industries in which the Japanese now excel—such as automobiles, consumer electronics, computers, and semiconductors—all originate in this country? This impression is a major cause of the frustration we now feel. If we are the inventors, how have the Japanese been so successful in adapting these inventions? Are our impressions really supported by the history of electronics?

These are important questions. Conventional wisdom suggests that leadership in the creation of the basic technology within an industry should lead to control of the industry itself. This apparently logical conclusion immediately runs into a problem: just what do we mean by "technology"?

For the moment, let's define technology as "the basic science, theories, and inventions that have been the roots of an industry." That's a good working definition most people would agree with. When we define technology in those terms, the notion that the United States has a monopoly on innovation in electronics does not stand up under critical examination. As we will see in this chapter, today's electronics industry grew out of many different disciplines. The bulk of the early development occurred in Western Europe, not the United States. With the invention of the transistor, the focus moved to the U.S. Today, the successful adaptation of these developments appears to be shifting more and more toward the Asian basin. To a large degree, these changes parallel the economic trends of the same geographic regions.

The Roots of Electronics

It would take a separate book to do justice to an examination of the history of electronics. For our purposes, however, we only need to look at those developments and people that have led us to where we are today.

Electricity was "discovered" when people first noticed that if amber (*elektron* in Greek) was rubbed with fur it would attract light objects such as lint. Thales of Miletus, a Greek philosopher, recorded this observation around 600 B.C. In effect, he had observed what we now refer to as "static cling"! More than 2,000 years later—in 1600—Queen Elizabeth's court physician, William Gilbert, began a systematic investigation of the electrification of materials such as amber. This was followed by the related work of Dufay in France, Bevis in England, America's Ben Franklin, and others. In 1785 Charles Coulomb of France developed a mathematical explanation for this phenomenon. With the exception of Franklin's famous experiments, Western Europe was the center of these studies. The lack of American contributions is not surprising, since the U.S. was just emerging as a new nation.

While these experiments and observations were interesting, they did nothing to put electricity to practical use. Electrostatically generated electricity is, by definition, an electric charge that tends to stay in one place. A major step forward happened in 1800 when Alessandro Volta of Italy developed the *voltaic pile*. This was the first electrochemical battery and a useful source of direct current (DC). Since direct current flows readily from one point to another through a conductive material (such as copper, aluminum, or silver), it became possible to make electricity do useful work. One of the first applications was in Humphrey Davy's electric arc lamp, developed in England in 1808.

The early 1800s were a particularly fruitful period in defining the interrelationship between magnetic phenomena and the flow of electric current. In 1820, Hans Christian Oersted of Denmark observed that electric current creates a magnetic field around it. In 1831, Joseph Henry of the United States and Michael Faraday of England used Oersted's discovery to lay the foundations for electric generators and motors. Faraday was a particularly prolific scientist whose contributions included his theory of electromagnetic waves, laws of electrolysis, investigations on the conductivity of silver sulfide, and introduction of the concept of dielectric polarization. A landmark contribution was made in 1864 by James Clerk Maxwell of Scotland and publication of his *Treatise on Electricity and Magnetism*. This elegantly united the interrelations between electricity and magnetism in four mathematical equations. The study of Maxwell's equations is a rite of passage for today's electrical engineering students.

Many people (especially in the United States) think of semiconductor technology as a uniquely American development. However, unusual phenomena in what would be known as semiconductors were first observed in the nineteenth century by European scientists. For example, Faraday reported the anomalous conductivity of silver sulfide in 1833. In France, Becquerel studied illuminated semiconductor/electrolyte interfaces in 1839. Willoughby Smith of England investigated the photoconductivity of selenium in 1873. The following year, Karl Braun of Germany reported experiments on the rectification of metal/lead sulfide junctions. These investigations were important precursors in the development of the transistor since they showed there were different ways of controlling the flow of current in certain types of solid materials.

While Europe was still the center of research activity, American inventors were proving adept at applying basic scientific concepts to the development of useful products. In 1837, Joseph Davenport developed the first commercial electric motor. Three years later, Samuel Morse introduced the electric telegraph. In 1875 Alexander Graham Bell invented the telephone. Thomas Edison invented the phonograph in 1877 and, two years later, applied for an incandescent light patent. In some ways, the United States was the Japan of the nineteenth century; it built a robust, modern economy on discoveries and innovations made elsewhere.

Many believe that modern electronics started with the so-called Edison Effect. In 1883, while trying to improve the life of his new light bulbs, Edison introduced a metal plate into the evacuated enclosure and observed that a current could flow to it from the heated filament. The effect was published and patented, but Edison did not attempt to commercialize it. Perhaps absorbed with his many other projects, he apparently did not recognize the potential importance of what was—in effect—a vacuum diode. Credit for giving the world the vacuum tube went to two other Americans, John A. Fleming for his 1904 diode experiments, and Lee de Forest for his Audion, or vacuum triode, which was introduced two years later. Coupled with the earlier 1887 work of Germany's Heinrich Hertz on radio waves and the wireless telegraph introduced by Italy's Guglielmo Marconi in 1896, these developments laid the foundations for radio. Equally important, the cathode ray tube (CRT) first appeared in 1897, based on the work of J. J. Thomson of England and Karl Braun of Germany, and made it possible to display images electronically.

As these practical forerunners of today's electronic systems began to appear, scientists continued their attempts to understand matter and, in particular, the nature of the electron. Once again, Europe was the center of research activity. In 1895, H. A. Lorentz of Holland postulated his theory of electromagnetic waves. Three years later, J. J. Thomson determined that the electron was the basic element of electricity. In 1911, England's Lord Rutherford developed a solar system-like model of the atom based on classical mechanics. (Little known is the fact that five years earlier Hantaro Nagaoka of Japan had published a similar version of atomic structure.)

Unfortunately, the "solar system" model could not explain a number of observed anomalies, such as atomic spectra. A new theory of matter was needed, and that turned out to be quantum theory. It was first postulated by Max Plank of Germany in 1900, expanded upon by Albert Einstein in Germany in 1905, and used by Denmark's Niels Bohr in 1913 to modify Rutherford's model. The new view of matter described by quantum mechanics allowed electrons to move around the atomic nucleus only in discrete orbits or energy states. The next major advance came in the early 1920s, when de Broglie of France postulated that the electron had both material and wave-like properties. Finally, in 1926, Erwin Schrödinger of Austria united all the previous theories of quantum mechanics into what is now known as wave mechanics.

Such discoveries in physics set the stage for the development of solid-state electronics. It would be only a matter of time before the first true solid-state amplifier, the transistor, was invented. The transistor would be an American invention, but it was built upon discoveries and research from Europe.

America Takes the Lead in Solid-State

When did the era of solid-state electronics begin? Perhaps it was in the mid-1920s, when Julius Lilienfeld of the United States conceived a copper sulfide solid-state amplifier. Although it proved to be impractical, it was one of the earliest such ideas in the field. In that same period, materials work was underway in the United States and Europe on semiconductors such as silicon (Si), lead sulfide, and iron pyrite. This work was driven by their use as radio and, later, as radar detectors. In the 1930s, theories of rectification were refined—primarily in Europe—and scientists at Bell Telephone Laboratories (abbreviated BTL, or, more commonly, Bell Labs) began serious investigations to develop a practical solid-state amplifier. Then came World War II, and around the world all scientific effort was redirected toward military applications.

In retrospect, World War II actually accelerated the development of solid-state electronics, especially in the United States. For example, there was extensive government-sponsored research and development on the principal microwave radar detector materials, silicon and germanium (Ge). Those two elements were later the principal materials from which transistors and integrated circuits were fabricated. There was also a drive for miniaturization in such military applications as the VT-proximity fuze. Finally, the first digital electronic computers, such as ENIAC and EDVAC, were developed toward the end of World War II. These were built entirely from vacuum tubes and mechanical relays, and had very poor reliability. This unreliability was a strong impetus to develop more reliable versions using solid-state technology.

The United States emerged from the war as the world's strongest economic power. Its land masses were unscathed, and its industrial base was intact. It had enormous natural resources, including self-sufficiency in petroleum, iron ore, food, coal, and other vital materials. These physical resources were matched by an unparalleled pool of scientists and engineers which was growing rapidly through a flood of talented immigrants, primarily from Europe. All the pieces were in place for world leadership in electronics to shift from Europe to the United States. The symbol of this rise of American dominance was the development of the transistor at Bell Telephone Laboratories.

While the transistor would have eventually been invented somewhere, Bell Labs was ideally positioned to be first. In late 1945, Mervin Kelly created the Solid State Research Laboratory at Bell Labs and staffed it with some of the finest talent available. It was well funded, and, despite its apparently broad charter, very well focused—limit investigations to germanium and silicon (the two best understood semiconductor materials) and develop a practical solid-state amplifier. Success came a little over two years later when, on December 23, 1947, John Bardeen and Walter H. Brattain demonstrated the first Ge point-contact transistor. One month later, William Shockley conceived a more practical device, the junction transistor. The first useful such device appeared in the early 1950s, and the three Bell scientists shared the 1956 Nobel Prize in Physics for invention of the transistor. It is fair to say that BTL was the father of solid-state electronics as we know it today through its inventions, its license and know-how agreements, and the new companies that sprang from it.

Although Bell Labs invented the transistor, the true commercialization of it and the proliferation of technologies and products it spawned were primarily the result of a multitude of entrepreneurial start-ups. The earliest companies were spin-offs from Bell Labs; eventually, these spin-offs themselves served as the incubators for other new firms. Most of the early focus of this activity was the area south of San Francisco fondly referred to as Silicon Valley. The first of these new companies was Shockley Semiconductor Laboratory, founded by the Nobel Laureate in 1954, and staffed by some of the brightest young minds available in this new industry. However, Shockley was a difficult boss and, in 1957, a group of employees that became known as the "traitorous eight"—led by Bob Noyce, Gordon Moore, and Jean Hoerni—left to form Fairchild Semiconductor Corporation. This company, sadly now gone, was to become the early giant of the semiconductor industry.

But why was the explosive growth of semiconductors through start-ups and new companies, rather than at Bell Labs—where it all began—or within other existing large electronic corporations? For example, the new technology particularly threatened vacuum tube manufacturers such as Sylvania, Raytheon, RCA, GE, and Westinghouse. Although all made important contributions to the semiconductor field, none are important competitive factors today in the industry. Perhaps the reason why is that the size of such established companies was simultaneously their greatest strength and their greatest weakness in semiconductor technology. While the laboratories of such large companies offered an ideal environment for carrying out basic science and the early development of new products, those companies were also slow moving and had oppressive bureaucracies. The young pioneers working at the frontiers of solid-state technology wanted to move quickly, and this seemed impossible in these large corporations. This desire to get things done rapidly is still a major motivating force even today in new high technology ventures. (However, this does not mean that all new start-up ventures automatically have an advantage over established companies when it comes to corporate cultures. Shockley Semiconductor is a case in point. While Shockley had it all when his new company was first formed, he offered his own unique brand of control, which many of his employees found worse than the big company environment they had left.)

Innovative forms of financing also fueled the American semiconductor industry. In the investment banking field, Arthur Rock of Hayden Stone helped Noyce and his team link with Sherman Fairchild of Fairchild Camera and Instrument Corporation, resulting in the creation of Fairchild Semiconductor. The new company became a division of Fairchild Camera and Instrument when the investor exercised its investment agreement option to purchase the start-up. (However, this organization structure eventually became untenable for the Fairchild start-up team and ten years later Noyce and others left to form Intel.)

The start-up rage accelerated in the 1960s with the creation of the venture capital (VC) industry. The venture capital industry made it possible for almost any team with a good idea to get funding, start a new company, and, if successful, to become—along with the investors—very rich. My cousin, Peter Sprague, was an early pioneer in this industry when—in the mid-1960s while he was in his twenties—he put together the financing to hire Bob Widler, David Talbert, and finally Charlie Spork away from Fairchild to create National Semiconductor. National became one of the giants of the U.S. merchant semiconductor industry. Through the vehicle of equity participation, Peter offered these geniuses the opportunity of almost instant wealth. This started the breakup of Fairchild and a Silicon Valley tradition: the diffusion of semiconductor technology through new company creation. To this day, Peter continues as National's chairman.

Using venture capital to launch a new company remains a uniquely American phenomenon. In countries such as Japan, venture capital is still very much an infant industry. Why is this so? With or without venture capital, starting a new company is one of the riskiest of all endeavors. For investors, venture capital financing is perhaps the riskiest of all investment approaches (but also offers the highest potential return). Yet venture capital has historically been available for new electronics companies. I believe this is due primarily to the fact that, of all nationalities, Americans are the greatest risk takers. This is largely due to our heritage as an immigrant society. Leaving home to seek opportunity in a new country is perhaps the greatest risk there is. Venture capital and the related vehicle of equity participation for the new management teams are probably the major reasons for the subsequent proliferation of new semiconductor companies.

How did northern California become the epicenter of the American semiconductor industry? While I am not sure why Shockley located his new company in what would become Silicon Valley, I feel the "risk factor" mentioned in the last paragraph was certainly an important reason why that locale appealed to such industry pioneers. The area seemed to be the last frontier in the United States and, in a way, the founders of the first semiconductor companies were like backwoodsmen moving west to seek their fortune. The atmosphere encouraged innovative, unconventional thinking as well as personal and career mobility. In those days, Silicon Valley was also one of the most beautiful places in the world. I will never forget my first view of the Bay Area as my wife and I drove over the top of the Oakland hills in the fall of 1953, when I had been selected to attend the U.S. Navy's Officers Electronics Material School at Treasure Island. Before us, under a low layer of clouds, spread a panorama that encompassed Oakland, Berkeley, the Bay Bridge, the Golden Gate Bridge, and the skyline of San Francisco. Then there was no smog, the hills were largely devoid of development, the highway system handled the traffic with relative ease, and it was love at first sight. I'm sure many others who helped launch the American semiconductor industry must have felt the same way.

While the transistor was important, it was the development of the semiconductor integrated circuit (IC) that transformed Silicon Valley forever. An IC is a single block of semiconductor material in which all of the elements of an electronic circuit have been created. However, the IC was not born in Silicon Valley but instead Silicon Prairie—Dallas, Texas, to be precise. Jack Kilby of Texas Instruments (TI) is generally credited with inventing the integrated circuit. In the fall of 1958, he demonstrated a phase shift oscillator integrated on a single chip of Ge crystal. It was a crude device, with jumper wires responsible for interconnecting the different circuit elements. Practicality came from Fairchild where, a year later, Jean Hoerni developed the planar process which made possible ICs as we know them today. The resulting transistors were superior in performance and the surface of the Si wafer was flat. This meant its circuit elements could be interconnected by metal runs, deposited using vacuum evaporation, instead of wires. The final pieces fell into place in 1960 at the International Electron Device Conference in Pittsburgh. Here Bell Labs presented separate papers on the epitaxial transistor and the metal-oxide semiconductor (MOS) transistor, while Fairchild announced the first family of planar integrated circuits, RTL (resistor-transistor logic) Micrologic. In many ways, the conference marked the birth of the IC industry and era.

Those were heady times, and I speak from personal experience. Although I really did not know it at the time, I was in the middle of the action. I entered Stanford University in 1955 to pursue a Ph.D. in physical chemistry, and it was entirely fortuitous that my thesis advisor, Claudio Alvarez ("Butch") Tostado, was one of the few chemists in the country at that time involved in the study of Si chemistry. My thesis, "The Nature of Metal-to-Semiconductor Alloy Junctions," was accepted by Stanford in the spring of 1959 but never published. Although certainly not Nobel Prize material, it was very timely and generated considerable initial interest at such places as Shockley Labs and Fairchild. Unfortunately, this interest quickly evaporated upon realization that I was the son of Robert C. ("R.C.") Sprague, the founder and president of Sprague Electric! I clearly recall a brief conversation with a slight, intense Bill Shockley who concluded, "I would be happy to help support your work at Stanford, but I am not about to let you in my labs." The Shockley defectors who launched Fairchild had made their point.

The same thing later happened to me in my second Fairchild interview when they also made the familial connection. But for my last name, I could have been involved with Silicon Valley from its start!

I ended up at Sprague Electric, working in their research labs on surface states in Si. It was in that capacity that I attended the 1960 Pittsburgh conference. Luckily, Sprague Electric was no slouch in semiconductors. My first boss, Kurt Lehovic, was an outstanding scientist who was awarded the basic patent on p-n junction isolation in 1962. While Silicon Valley was the center of the early semiconductor industry, it did not have a monopoly on talented, creative people and innovative companies.

What About Japan?

So the roots of solid-state electronics began in Europe and the focus moved to the United States after World War II. Where were the Japanese all this time? While a history of solid-state electronics written in Japan might be more generous, it is not unfair to say the Japanese made precious few fundamental contributions to the field.

There are a number of reasons for this. Until World War II, Japan had operated primarily in the Asian sphere. As a geographically remote island nation, they had remained largely insulated from outside influences from Europe and the United States. With a few notable exceptions, there were only a handful of exceptional scientific universities compared to the United States and Europe. There were no corporate equivalents to Bell Labs. Language was also a problem since most of the scientific literature was in English, German, and French. While at least a reading knowledge of English or another European language is common among Japanese researchers today, few Japanese had such skills prior to World War II. The state of pre-World War II Japanese research was vividly described in a special report on "Scientific Research & Education in Japan" that appeared in the December 2, 1991 issue of *Chemical & Engineering News* (C&EN). In it, Professor Koji Nakanishi of Columbia University and the Suntory Institute for Bioorganic Research in Osaka concludes that until after World War II the primary focus of university research in Japan was on catching up with the West.

Following the war, Japan had to regrow a scientific community almost from scratch. Its industrial and social infrastructure had been—literally—blown to bits. Much of the cream of its youth had perished in the war. In addition, the flood of scientific talent that left Europe before and after World War II went to the United States, not to Japan. In light of this weak starting base, what the Japanese have been able to accomplish so quickly in semiconductors is genuinely extraordinary.

Following World War II, the Japanese did start to make several important early contributions to the field. For example, during the 1950s Jun-ichi Nishizawa introduced the concepts of the PIN diode and static induction transistor, and also filed a patent application on ion implantation. Even more importantly, in 1957—a year before the work of Kilby and several years before Noyce—Yasuo Tarui filed a patent on integration of a bipolar and a field effect transistor (FET) on a single chip. In effect, this was an integrated circuit. However, Tarui would never receive credit for this invention because the patent was specific on only this combination and never reduced to practice. The tunnel diode was invented by Leo Esaki in 1958. And in 1960 M. Kikuchi introduced concepts that, in 1963, would lead to invention of the Gunn diode by J. Gunn of IBM.

How did the Japanese recover so quickly from the war's devastation to make such contributions? One important factor was clearly the major economic assistance provided by the United States, which allowed the Japanese to rebuild their economy from the ground up. As a country almost devoid of natural resources and prohibited from a military capability, Japan had to create an industrial foundation by adding value through the skills and determination of its people. While capital was initially scarce, it began to quickly accumulate thanks to the extraordinarily high savings rate of its citizens. This was partly cultural, partly the result of government tax policies and the absence of any real social safety net, and partly due to the fact that—at least initially—there was little else to do with the funds.

Like the rest of the world, the Japanese quickly realized the future potential of semiconductor-based electronics. As soon as possible, certain Japanese companies negotiated license and know-how agreements with Bell Labs. In 1953 and 1954, Kobe Kogyo—later part of Fujitsu—and Tokyo Tsushin Kogyo—now Sony—were among the first to do so. With military markets blocked and a large potential home market accumulating capital, consumer electronics seemed the logical first step. The first consumer application was a Sony transistor radio introduced in 1955. This began the Japanese consumer electronics industry that today is dominant worldwide. The first Japanese ICs did not appear until 1965 and were manufactured by NEC. Again, the application was consumer, specifically an IC-based calculator introduced by Sharp in 1966. In contrast, by this time there were already more that 25 American companies manufacturing integrated circuits when the Japanese introduced their first ICs.

The Japanese recognized that, to afford the natural resources, goods, and services they needed to purchase from the West, they needed to develop sources of foreign capital and this could only be accomplished through dynamic export trade. Stung by the world perception that Japanese (and, in fact, all Asian goods) were basically shoddy, they eagerly sought help from the United States. In 1950, W. Edwards Deming was invited to give a seminar in Japan on statistical process control. Four years later, J. M. Juran began to introduce quality management techniques into Japanese corporations. Culturally precise, and seeing the enormous potential of adding value through quality, it became both a national passion and agenda. The Japanese took the lessons of Deming and Juran to heart. Although American firms have made great strides in recent years, quality continues to be a catch-up game for much of the U.S. semiconductor industry. The student has now become the teacher.

The Japanese emphasis on the consumer market was in contrast to the United States semiconductor industry. In the United States, the computer industry was the principal driving force and the government was the industry's largest customer right into the 1970s. While today we lament the large percentage of U.S. research and development supported by the federal government and the poor record of making it useful in the industrial sector, such funding was vitally important to the infant U.S. semiconductor industry. The most important early applications were, in fact, military. The Minuteman missile program of the 1960s was the first major user of ICs as well as many other types of components. Quality was assured through elaborate test systems and measured on an AQL (acceptable quality level) basis. Component costs were also very high. This began to change in the mid-1960s with introduction of Digital Equipment Corporation's PDP-8 minicomputer and the IBM System 360 computers, the first high volume user of semiconductor-based circuits.

The early orientation of Japanese on consumer applications provided a very different environment. With quality as a national priority, semiconductor companies were forced to build quality components at low cost in increasingly large volumes. Using the teachings of Deming and Juran, they strove to build quality in instead of testing defects out. They came to embrace the Boston Consulting Group concepts of learning curve costs, pricing, and growth. Market share became the driving force in their industrial growth. Even with only a few early competitors, in some ways Japanese component suppliers faced a tougher environment than their U.S. counterparts. However, since consumer electronics was considered "simple" and military/industrial "difficult," U.S. semiconductor firms tended to feel the Japanese would never be a competitive threat in semiconductors, particularly with their very late start in integrated circuits. How wrong that assessment proved to be! Using their largely unchallenged and captive consumer electronics market, Japanese component suppliers perfected their technologies, improved their quality, and drove down their costs. Using this as their base, by the 1970s they were well-positioned to attack the industrial market, particularly with memory ICs for the computer industry, which was largely based in the United States at that time.

U.S./Japan Semiconductor Competition Heats Up

The technology explosion of the 1960s continued unabated with the United States initially leading the way. RCA developed metal-oxide silicon (MOS) technology, which had low power consumption but was somewhat slow. As photolithography improved, MOS and its many variations evolved as the dominant technology when memory and microprocessor revenues became the driving engines of the 1970s. The mid-1960s also saw the rise of linear circuits. In 1964, Zenith developed a hearing aid with an IC audio amplifier, the first major U.S. non-military IC application. Bob Widler developed the μA 702 operational amplifier at Fairchild in 1965. This was the predecessor to a family of increasingly sophisticated operational amplifiers, including Widler's μA 709 and the LM 101 he developed after moving on to National Semiconductor. Fairchild and others began to introduce ICs specifically designed for consumer applications, a niche pursued by Sprague Electric in the late 1960s after Dr. Robert Pepper joined them from the University of California at Berkeley. Pepper's highly competent design team developed sound circuits for Delco automotive radio applications, color television circuits for Zenith, and flashbar ICs for Polaroid's SX-70 camera.

It was in the late 1960s when American IC companies first began to experience competition from the Japanese. Sprague first felt such competition in the integrated circuits it supplied to Zenith for use in their solid-state television sets. Our competition was Hitachi, which was committing a much larger engineering team to Zenith than Sprague could muster. Before long, we were faced with losing the bulk of the business. During a review meeting at Zenith, I tried every argument I could marshal. I pointed out what a good job Sprague Electric had done for them over the years, first in capacitors and now also in semiconductors. I argued that, as a small specialist, the Sprague Semiconductor Division could offer superior customer service since consumer linear ICs was the market niche we had chosen to concentrate on. Most importantly, I tried to convince them—to no avail—that what they were really doing was help create another strong competitor for Zenith itself. Price and quality were never issues, only the size of the engineering resources Hitachi was committing.

We eventually lost the argument—we could match Hitachi on price and quality, but not size—and the account. But, as I had predicted, Hitachi became one of the leading world television competitors. Now Zenith is the sole U.S. survivor of the television industry, and only barely, thanks to competition from Japanese companies like Hitachi. Sprague's experience was a forerunner of what was to come for other U.S. semiconductor firms in other market niches. Our only recourse was to seek other market niches. The power interface ICs that Sanken was selling for Sprague in Japan were one of the new niches Sprague entered as a result of this setback.

Most American semiconductor companies in the heady 1960s were like Zenith in consumer electronics—their attention was not on the possibility of Japanese competition in the future. Fueled by readily available venture capital, and spurred by the success of Peter Sprague's raid on Fairchild, Silicon Valley became an incubator for new semiconductor start-ups. In 1966 and 1967, there were just three major ones. But in 1968 there were thirteen, followed by eight more in 1969.

One of the most successful was Intel, formed in 1968 when a frustrated Bob Noyce, Gordon Moore, and others finally left Fairchild. Intel was the early leader in both static random access (SRAM) and dynamic random access (DRAM) memories. Intel's most important contribution occurred in 1969 when a design team, lead by Ted Hoff, conceived the first single chip central processor unit (CPU) or *microprocessor*. This was introduced to the market in 1971 as the 4-bit 4004 microprocessor, and was followed by the enormously successful 8-bit 8008. Incredibly powerful computer power became available at unbelievably low prices and the electronics industry would never be the same. Today, Intel remains the world leader in microprocessors as a result of their selection for use in IBM and IBM-compatible personal computers. The United States microprocessor industry still enjoys a commanding lead over that in Japan and Europe. It is therefore ironic that the 4004 was initially developed by Intel as part of a desk top calculator project for a Japanese corporation, Busicom. While the microprocessor was clearly an American innovation, its development was driven by a Japanese company!

As Silicon Valley boomed, the semiconductor industry was heating up in Japan as well. Recognizing that ICs would form the linchpin of all electronics in the future, MITI (the Ministry of International Trade and Industry) in 1966 introduced the first of a series of IC-related initiatives, known as the IC Product Strategy for Japan. While not terribly successful, it was the first attempt to place some rationale to the IC industry and Japan's position in it. As was true with later such initiatives, the primary goal was to make the Japanese computer industry more competitive. At the same time, increasing numbers of talented Japanese engineers and scientists began to attend U.S. universities and to work in this country. One excellent example is Masatoshi Shima, a key figure at Busicom during the Intel calculator project and development of the first microprocessor. He joined Intel in the U.S. in 1972, went on to Zilog to design the famed Z80 microprocessor, and eventually returned to Japan in 1979 to set-up Intel's Japanese design center.

Despite Japan's efforts, the U.S. entered the 1970s as the world leader in semiconductors and would maintain that position until the mid-1980s. Start-ups and spin-offs continued to proliferate. One of the more important was Mostek, formed in 1969 when most of the Texas Instruments MOS team, led by L. J. Sevin, decided to go their own way rather than move from Dallas to Houston as "requested" by TI management. Mostek was one of the early leaders in semiconductor memory ICs. Most of the early equity came from Sprague Electric, and I was the plant manager at Sprague's Worcester, Massachusetts plant where Mostek did its first manufacturing. They were also the first to use ion-implantation, developed by Dr. Ken Manchester in the Sprague research and development labs, as a manufacturing process. Mostek was also involved in another Busicom calculator project, this time for a hand-held unit. Because memory ICs were its principal products, Mostek was one of the first American companies to experience the Japanese competitive threat in semiconductors head-on.

By the mid-1970s, highly reliable Japanese memory ICs were beginning to penetrate the U.S. market and compete against those from companies like Mostek. In 1976, MITI created the much-heralded VLSI (Very Large Scale Integration) project involving Fujitsu, Hitachi, Mitsubishi, NEC, and Toshiba. The goal of that project was to make the Japanese computer industry competitive with IBM by the early 1980s. The trick for MITI was to get those five fiercely competitive Japanese corporations to work together to develop the necessary technology. The project continued for four years with a total budget of approximately $200,000,000, of which 60% came from the five companies. Today, there is a legitimate argument over how successful the project actually was compared to what each company could have accomplished alone. Still, the results were impressive: some 1000 related patents were issued and close to 500 papers published in such important areas as VLSI design, processing, lithography, testing, and devices.

Despite such Japanese accomplishments in the mid- to late-1970s, most U.S. semiconductor companies still considered their Asian counterparts to be only copycats. However, knowledgeable U.S. executives who actually visited Japan began to express concern. I remember well a 1977 Mostek board of directors meeting at which Sevin, having just returned from a lengthy tour of Japanese semiconductor firms, grimly told us he felt the Japanese would eventually bury the U.S. in the DRAM business. Unfortunately, he was right. Within two years, a Mostek battered by Japanese competition would be bought by United Technologies. Soon after the sale, most of the key management left. Even the deep pockets of United Technologies couldn't save Mostek; by the end of the 1980s, it would disappear as a corporate entity. (An aside: L. J. Sevin left the semiconductor business after Mostek and is today a venture capitalist, helping others secure the same sort of funding that launched Mostek. He says that while the venture capital business lacks the excitement of running a semiconductor company, "it ain't all bad.")

Japan's copycat image began to change in 1978 when Hitachi stunned the industry by introducing a 4K CMOS SRAM IC that was superior in performance to any of its U.S. competitors. It had a considerably smaller die size as well. This device was developed by Toshiaki Masuhara, who received his education at the University of Kyoto and the University of California at Berkeley. As the 1980s progressed, Japanese firms introduced a steady flow of DRAM and SRAM devices that were superior to those from American companies in terms of cost, quality, and/or performance. One by one, U.S. firms dropped out of the memory IC business throughout the 1980s. Today, the Japanese absolutely dominate the non-captive DRAM market, although captive producers, like IBM, still produce substantial quantities of memory devices in the United States. As mentioned earlier, the United States continues to enjoy a comfortable world lead in microprocessors and related microcontroller and peripheral chips.

The current relative market shares of the competitors in these different segments provides real insight into the strengths and weaknesses of the two countries. Success in DRAMs is driven by processing and manufacturing skills as well as the availability of large amounts of capital—all areas where the Japanese have competitive advantages. However, with microprocessors and related devices, the U.S. skill in design is the key element. This is also true of application specific IC (ASIC) devices, felt by many to be the future of the industry. This is largely the result of Lynn Conway and Carver Mead's seminal book *Introduction to VLSI Systems*, published in 1979 as an outgrowth of a course they taught at the California Institute of Technology. In this book the authors bring together computer science and electrical engineering and teach a systematic method of synthesizing complete systems on a single chip of silicon or, in other words, the creation of true VLSI.

It is almost impossible to hold an objective discussion concerning the relative contributions of the United States and Japan in the semiconductor market. You can support almost any argument you want by carefully selecting which experts you cite. One example is from the 1984 book *The Competitive Edge: The Semiconductor Industry in the U.S. and Japan*. The authors of chapter two—Michiyuki Uenohara and Takuo Sugano of Japan and John Linvill and Franklin Weinstein of the United States—differ markedly on the relative contributions of the two nations to the industry. The Japanese authors argue their country has made major contributions to both new manufacturing technologies and to market development, while the Americans counter that the Japanese have been primarily followers of U.S. companies in both. All four do agree that the Japanese contribution to basic research and new semiconductor technologies has been very limited, at least up to the early 1980s. By contrast, George Gilder in his book *Microcosm* argues that intellectual capital is far more important than either financial or physical capital, and that the United States far outstrips the rest of the world in "brain power" capital.

In a letter critiquing the initial proposal for this book, one reviewer stated flatly that the United States had not lost the entire semiconductor market, but only the memory end of it. Perhaps. Nevertheless, at one time the U.S. owned the entire market but has consistently underestimated foreign competition, especially the Japanese. Statistics support this conclusion and present a picture that can give us little solace. Although the numbers differ depending on the source, a conservative estimate of market share in 1989 had Japanese firms accounting for 38% of the non-captive world IC revenues of $40,000,000,000 versus 32.5% for the United States. This difference was largely driven by sales of memory ICs. This disparity is reflected in the ranking of the world's largest semiconductor firms. In decreasing order of worldwide semiconductor sales, the top suppliers were NEC, Toshiba, Hitachi, Motorola, Texas Instruments, Mitsubishi, Fujitsu, Intel, Matsushita, and Philips/Signetics. It must be pointed out that these numbers do not include the captive production figures of a number of U.S. companies, including AT&T and IBM. The latter is probably the largest manufacturer and consumer of ICs in the world.

Regardless of which experts one wants to believe or how statistics are interpreted, one cannot avoid the conclusion that, without a position in the early technology or an apparent bent toward entrepreneurship, Japanese companies have forged an increasingly dominant position in ICs, an industry that unquestionably is the driving engine in all of electronics.

The Role and History of Capacitors in Electronics

While ICs are glamorous and the "engines" of contemporary electronics, they still make up a relatively small percentage of the total dollar value in a huge industry. For example, in 1987 the largest parts of the $223,000,000,000 U.S. electronics industry were the communications (27.2%) and computer/industrial (32.9%) end equipment segments. Electronic components were only 18% of the market. Of this 18% (or $40,000,000,000), $16,000,000,000 were semiconductor devices, of which integrated circuits were $12,000,000,000. Other components include capacitors, vacuum tubes, resistors, inductors, connectors, wire and cable, and a number of other categories. As mentioned in the previous section, in 1989 ICs accounted for 80% of a total $50,000,000,000 world semiconductor industry. The comparative number for capacitors was only $8,000,000,000, of which $3,500,000,000 was from Japan and a stagnant $1,500,000,000 from the U.S. The remainder came primarily from Western Europe. Given these comparatively small portions of the total electronics market, it's appropriate to ask just where components such as capacitors fit into our chronology and if they are really important in electronic systems.

When the IC was invented, it was predicted that the bulk of passive components such as capacitors would quickly disappear. This is because the capacitive function can be incorporated in an integrated circuit by using either a p-n junction or an MOS structure. Circuit techniques can also be used to dramatically reduce the requirement for both discrete resistors and capacitors—up to a point. Capacitors integrated on a chip are very limited in their charge storage capability, and also take up considerable surface area. This increases the size of the IC chip. For such reasons— although capacitor revenues have decreased as a percentage of total electronic components—the absolute numbers have continued to increase, driven by the explosive overall growth of electronics. Consumer products such as color television and compact disc players can use hundreds of capacitors per system. Computers, depending on capacity and function, can use many thousands. However, the mix has been changing. In the early days, paper capacitors dominated. Next aluminum electrolytic types dominated. Today, because solid-state circuits operate at low voltages and the drive is for the maximum volumetric efficiency, the predominant family is the multilayer ceramic capacitor (MLC). MLCs now account for close to 40% of the world dollar volume and 80% of the units. This shift is forecast to continue.

Conceptually, a capacitor is about as simple a component as one can imagine. It consists of basically an insulator sandwiched between two conductors, and is used to store electric charge as an electrostatic field. The first condenser (the early name for the capacitor) was the Leyden Jar. It was first conceived independently in Germany and in Holland (why else would it be called a Leyden Jar?) in 1745, and perfected by John Bevis in England a year later. The first application was to store electrostatic energy generated by an "electrification machine." Ben Franklin used them in cascade in his famous lightning experiment. A Leyden Jar consisted of a glass jar (an insulator) surrounded by two pressed metal foil conductors. The conductors were placed so that they could not touch each other and produce a short circuit. Today's capacitors have different types of insulators, such as paper, Mylar, electrolytic compounds, mica, ceramics, or even air. This insulating material is known as a *dielectric*, and the ability of a capacitor to store electric charge is related to the type of dielectric used. The two conductors, called *plates*, are separated from each other by the dielectric, and are connected to voltage sources of opposite polarity.

Capacitors block the flow of direct current (DC) due to the insulating dielectric. If DC is applied to the conductors in a capacitor, electric charge will be stored as an electrostatic field until the capacitor's limit is reached. However, alternating current (AC) can flow through a capacitor as each conductor swings from positive to negative and back. This ability to block the flow of DC while permitting AC to pass accounts for the bulk of many capacitor applications, such as power supply filtering, motor starting, the transfer of electrical signals from one section of an electronic circuit to another, and—accounting for much of its enormous use in solid-state circuitry—to decouple or eliminate unwanted electrical noise.

The ability to store electrical charge, or capacitance (C), of a capacitor is directly proportional to the area (A) of the device and the dielectric permitivity (K) of the insulator compared to that of free space (K_0). K is a unique property for each different material and measures the ability to polarize the insulator. C is also inversely proportional to the thickness of the dielectric. The resulting formula for a parallel plate capacitor is the following:

$$C = K_0 K A / d$$

where $Ko = 0.225 \times 10^{-12}$ Farads (F)/in,
K = a dimensionless property of the insulator,
A = the area (in.2),
d = thickness (in.).

Capacitance is expressed in Farads (F), or more usually microfarads (10^{-6} x F), after Michael Faraday for his systematic studies on the polarization of dielectrics. The amount of charge, Q (expressed in coulombs) that such a capacitor can store is CV, where V is the applied voltage.

Other important characteristics relate to the energy loss of the component, the breakdown strength and resistivity of the dielectric, temperature coefficient, and failure mechanisms. As we have previously pointed out, the drive for high volumetric efficiency has been one of the principal reasons for the ever increasing importance of MLCs.

The history of capacitors provides an interesting parallel to the evolution of solid-state circuitry. The use of capacitors on a commercial scale did not occur until the latter 1800s, first in the telegraph, then the telephone, and finally in radio. While all this was going on, the properties of different dielectric systems were being systematically investigated. Because of the thickness of the glass and low dielectric constant, the Leyden Jar had a very low capacitance. In 1845, the use of stacked layers of thin, cleaved, natural mica interleaved with metal foil electrodes demonstrated the possibility of much higher capacitance. This was due to the thinness of the mica layers and the much greater area created by the alternating layer construction. This, in effect, was the first MLC. In 1854, Sir Charles Wheatstone of England noted the rectification properties of an aluminum (Al) electrolyte interface. This was the predecessor of the important Al electrolytic capacitor. In 1876 constructions involving a wound section of alternating paper and metal foil, impregnated with paraffin, appeared. This basic construction is still used today in some types of paper capacitors. Then, in 1896, C. Pollack was issued an English patent on a "wet" electrolytic in which a very thin layer of alumina (Al_2O_3) is grown by anodic oxidation on a piece of aluminum foil. This construction and its many variations today accounts for a little over one-third of world capacitor revenues. Capacitance is increased by etching the aluminum foil to create more surface area. Performance was improved when, in the mid-1920s, Sam Ruben, an independent U.S. inventor, filed the basic patent on the "dry" electrolytic which uses a non-aqueous electrolyte. He also proposed the use of tantalum (Ta) as an alternative to Al. Ta has a number of advantages, including an oxide with a larger dielectric constant (26 for the tantalum oxide dielectric versus 8 for alumina). Even higher capacitances can be obtained by using a sintered pellet of high purity Ta metal. The principal disadvantage lies in the high cost of the tantalum starting material.

The thrust in the 1930s in multilayer constructions was to replace natural mica by synthetic dielectrics, the first being vitreous enamel, then titania, and then barium titanate ($BaTiO_3$). Depending on the specific titanate formulation, Ks in the thousands can be realized. As with semiconductors, World War II also provided major impetus in capacitor development. The driving forces were miniaturization and the search by the military for storage elements that did not leak. Along these lines the U. S. Signal Corps supported three different approaches to manufacture MLCs: sprayed vitreous enamel dielectric layers (DuPont); stacked glass sheets (Corning); and stacked tapes of vitreous enamel (MIT). Using a modification of the DuPont equipment, Jack Fabricius and George Olsen of Sprague Electric in 1958 announced the first production MLC capacitor based on barium titanate. This can be considered as the "birth" of the modern MLC. Figure 2-1 is a cutaway of a conformally coated MLC capacitor and shows the multilayer construction. Figure 2-2 shows how such a design increases the area of the device and therefore the capacitance. Imagine a long strip of ceramic material on either side of which are deposited offsetting metal strips (the offset is necessary for the later termination and lead attachment steps). If such a construction could be folded back on itself like an accordion, one would have the same area but in a more compact block of material. The MLC build-up process accomplishes the same thing.

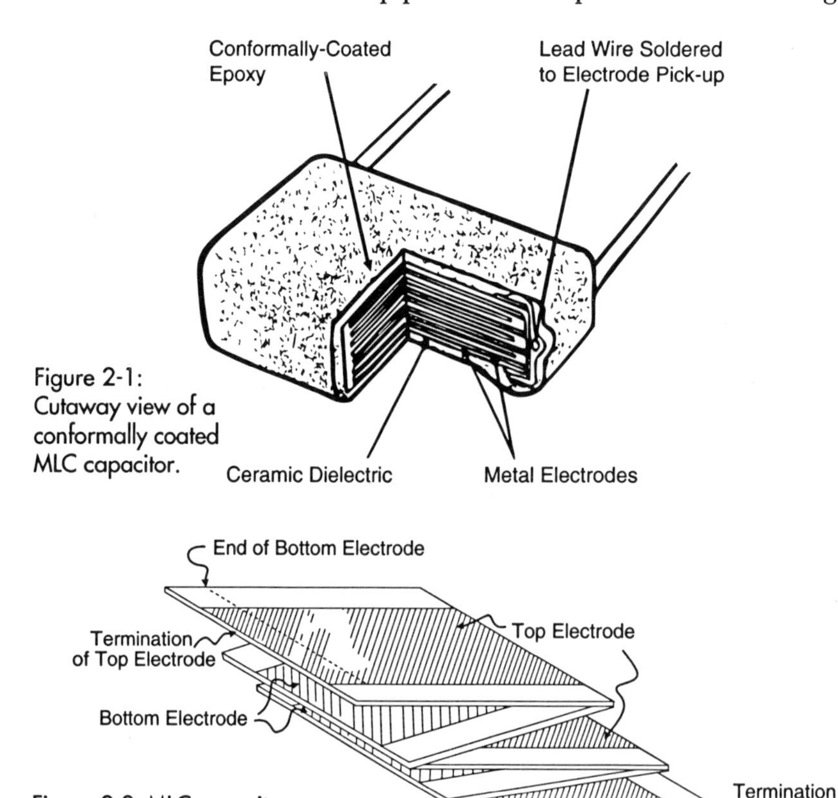

Figure 2-1: Cutaway view of a conformally coated MLC capacitor.

Figure 2-2: MLC capacitor viewed as an accordian structure.

There was one more major development to come. It had been postulated years before by Ruben, but seemed to have nowhere to go because of cost. In 1951, Dr. Preston Robinson, Sprague Electric's most prolific scientist, filed what was to be the controlling patent on solid electrolytic capacitors. The construction described by Robinson involved an aluminum anode and a lead peroxide solid counter electrode or cathode. In 1956, Taylor and Haring of Bell Labs published what would be the prevailing construction, a sintered pellet of high purity Ta powder which was anodized to form the dielectric and, most important, which had a solid cathode of semiconducting MnO_2. After a bitter court battle (which has left scars to this day), Robinson triumphed and his patent issued in 1962.

However, it was a combination of the invention of the transistor, which requires low operating voltages, and military requirements that made the solid Ta capacitors a commercial reality. Since high cost was more than compensated by the need for high volumetric efficiency and non-leaky electrolytes, this type of capacitor ideally fitted the bill for military systems. As with integrated circuits, the Minuteman missile was the first major application of solid tantalums.

The technology pattern in capacitors is eerily similar to semiconductors. Early development and technology came from Europe and then was successfully adapted and expanded in the U.S. This trend was further accelerated by the semiconductor revolution and the needs of the U.S. military. Led by Sprague Electric, the U.S. dominated the worldwide industry into the 1970s. Early imports from Asia to the U.S. were of poor quality. The major concern was dumping of such components as miniature aluminums, which were first used by the Japanese in consumer applications. As in semiconductors, however, the Japanese learned quickly. Today, the picture is very different and the capacitor world is increasingly dominated by Asian firms. Excepting certain special niches, American firms (Kemet and Sprague) are important world players only in tantalum capacitors. Aluminum capacitors are now primarily sourced by the Japanese, as are ceramic capacitors, including MLCs. This is due to the 1989 acquisition of AVX (by far the largest U.S.-based MLC supplier) by Kyocera. As a result, in the U.S. market Kyocera-AVX enjoys about 40% of the MLC revenues, Murata-Erie (Kyoto-based Murata is the world's largest MLC supplier) 15%, Kemet 15%, Vitramon (now owned by Thomas and Betts) 7%, and a host of smaller firms generally under 5% each. The early technology leader, Sprague, dropped out entirely during 1990, ending a battle that lasted more than thirty years to simultaneously be a major player in semiconductors, ceramic capacitors, and tantalums.

The story is similar in ceramic packages used for semiconductor encapsulation. Kycocera dominates this market, which is particularly important in military applications, and, along with two other Japanese giants, NTK and Narumi, accounts for over 95% of the world's production. Today, the Japanese lead the world in ceramic technology both for structural and for electronic applications.

This Japanese dominance of the capacitor and ceramics markets, like their domination of the memory IC business, is no accident. Japanese firms recognize that electronic components are important assets in their clear goal to dominate the world electronics market. They also recognize that ceramics are critical to many markets other than electronics. The process of capacitor and ceramic manufacturing is well-suited to many of the strongest advantages of Japanese culture. Capacitor and ceramic manufacturing also has parallels to many other activities, some of them which have been around for centuries. These points are the subject of our next chapter.

CHAPTER 3

The Process

Manufacturing ceramic capacitors is a fairly recent development in human history. But the basic process involved dates back to the dawn of civilization. This was made clear to me in the mid-1980s when my daughter, Cathy, was living in the Boston area and decided to take some courses in pottery.

Cathy studied a technique called "hand building," which is the formation by hand of clay into the desired ceramic shapes without using mechanical devices such as a potters wheel. While in school she worked with three different types of clay, known as earthenware, stoneware, and porcelain. This struck a familiar chord with me; in capacitor manufacturing, we call such different types of ceramic material *formulations.*

As in electronics, the different classifications vary primarily in the firing temperatures used. According to Cathy, clay is first excavated from the ground, dried, and treated to remove objects such as pebbles and rocks. Water is then added to give it the proper plasticity, and often a small amount of material referred to as "grog" is mixed in. Grog is finely ground-up pieces of fired clay and is especially important in facilitating the drying process of thick unfired ceramic pieces, or "greenware." Cathy once described her clay to me as "clean mud."

Once the resulting mixture has been formed into a desired shape, it is then placed in a furnace to be dried (or *fired,* as it's called). While electric kilns are common today, some potters still use simple pit kilns like those used centuries ago. Cathy was one of them, and soon we had one in the backyard of our home. Each time she drove up in her car she would bring at least one plastic bag filled with low-firing earthenware clay, the only material she could use in such a simple furnace. She also attempted to use clay taken directly from around the foundation of a new addition to our house. If necessary, she would mix in additional water to act as both a solvent and dispersant for the very fine clay particles, and soon various hand formed shapes were being fired.

It was fascinating to watch as she struggled with the different variables involved, as many of the steps were similar to what we did at

Sprague to manufacture MLCs. Before the clay could be used, it had to go through a process called "wedging," which is kneading the clay so that it is thoroughly mixed, dense, and free of air pockets. Failure to do so generally led to cracking during firing. If there was too much water in the clay, the unfired body wouldn't keep its shape; too little, and the shape couldn't be formed at all. Cracking could also be caused by improper control of the firing cycle.

In effect, Cathy was using the same process originally practiced around 5000 B.C. by potters in the New World, Southwest Asia, and China. On the other hand, she was also dealing with many of the same variables Sprague Electric encountered in electronic ceramics:

- different formulations
- control of additives, such as solvents and dispersants
- densification of the unfired body prior to sintering
- variations in furnace temperature and firing cycles
- control of shrinkage during firing

Today's technologists marvel at the skills of ancient ceramic artisans, which were developed without the benefit of modern science and analytical instrumentation. For example, only recently have the results of scientific investigation given Chinese ceramic scientists the ability to create replicas of Lonquan celadons that even fool the experts. (Lonquan celadons were exquisite, jade-like wares first developed by Chinese artisans in the eleventh and twelfth centuries.) One would think the expertise developed over the centuries by ceramic artisans, coupled with the enormous body of materials science now available, would yield an electronic ceramics technology that is a true science and not primarily know-how. However, that's not the case (although great progress has been made in recent years). When making both capacitors and pottery, much art still remains because of the many variables involved.

For the moment, Cathy's dismantled kiln and bags of clay are in the corner of our garage while she pursues a writing career in Albuquerque, New Mexico. However, I know she will return to her pottery someday. Working with ceramics seems to be in the family genes.

Ceramics at Sprague

The Sprague Electric Marshall Street complex in North Adams, Massachusetts was built in the late 1800s as a textile mill. It was purchased by Sprague after World War II for approximately $1,000,000. It consists of more than two dozen buildings covering close to a million square feet. These buildings were interconnected by a remarkable labyrinth of skyways and tunnels; one can literally get lost going from one part of the facility to another. My office as president was on the second floor of Building 4 in the southeast corner of the complex. While the view suffered, the wood-paneled office was large and pleasant, exuding the proper conservatism of a New England company founded in the 1920s. On the opposite end of the complex, the length of several football fields away and at the very northwest corner, was a basement with a more spartan atmosphere. Here the ceramics group was housed for many years. The contrast with my office was striking. Except for the encrusted windows, the spaces in that basement were relatively clean. However, the floor was of sealed concrete and the walls bare. Great kilns poured out thermal energy all year long. In the summer, the heat could be unbearable. There always seemed to be a slightly reddish haze in the air, the result of the ceramic formulations that were manufactured there. And, as I've previously noted, the steps they followed were much like those Cathy would follow years later in making her pottery.

There are many ways to fabricate ceramic capacitors. Regardless of the approach, one of the most critical procedures is control of the starting ceramic formulations. From the 1960s through the middle 1980s, this was one of the major responsibilities of Tom Prokopowicz, John Newman, and their small group of technicians at the Marshall Street complex. The material check-out procedure they practiced is an excellent example of some of the art that still exists today in electronic ceramics.

The major ingredient of most capacitor formulations is a ferroelectric material called barium titanate ($BaTiO_3$). Because of their internal crystalline structure, ferroelectrics contain spontaneously aligned electric charges and, as a result, can have large dielectric constants (Ks) which also peak at critical temperatures called Curie points. Until recently, when chemically-synthesized fine powders began limited use, the titanate was invariably made from barium carbonate ($BaCO_3$) and titania (TiO_2) using a high temperature solid state reaction. The most difficult compound to control is titania due to the variability of the inexpensive pigment grade material used by many U.S. manufacturers. (It is interesting to note that it was the Japanese who first offered a grade of TiO_2 specifically designed for electronic applications.)

For a number of years, Sprague used about 200,000 pounds of $BaTiO_3$ a year in its ceramic capacitor lines. It was first used in disc ceramics and then in MLCs. For cost reasons, Sprague purchased its raw materials in 100,000 pound lots and maintained a six month check-out and qualification cycle for releasing each lot to production. This cycle consisted of the following steps:

- Approximately 75 pounds of the starting materials (that is, $BaCO_3$ and TiO_2) were blended in water in what is called a homomixer to disperse the constituents.
- The material was then dried to remove the water and calcined (fired) in a furnace at about 1100° C.
- The calcined powder was then granulated through a sieve, and, along with the minor added constituents used to create the final formulation, ball-milled to reach the desired particle size.
- The resultant powder was pressed to form a disc, which was then fired and electroded to make a disc capacitor for electrical test.
- Tests included dielectric constant, loss angle, leakage current, temperature coefficient of capacitance (TCC), and life test under applied electrical load. Minor variations from specification could be adjusted by "tweaking" the final formulation. In non-scientific terms, this means adding "a touch of this and a pinch of that."
- Problems with load life could require using a completely new lot of titania and repeating the entire cycle.
- This process was repeated again and again until the formulation finally met specification and could be released to production. Production volumes were then shipped to the different ceramic manufacturing locations including Grafton, Wisconsin (where disc capacitors were made), Sprague's MLC plant in Wichita Falls, Texas, and Renaix, Belgium.
- In the meantime, the next lot of raw materials was usually in-house and the entire process began all over again.

For many years, starting materials were relatively impure and inexpensive—a few dollars for the carbonate and less than a dollar for titania. Not unexpectedly, about the only thing one could be sure of on a lot-to-lot basis was inconsistency. However, as the art progressed, purity improved, particle size and distribution specifications tightened, and the performance and reliability of the end product showed steady improvement. Recently, use of expensive, chemically synthesized $BaTiO_3$ has led to excellent reproducibility as new lots of material are received. Still, unlike the semiconductor world, we are not dealing with precisely doped single crystals and variations in materials do occur. Procedures similar to those I described above are still practiced throughout the industry. In other words, before the all important starting ceramic materials can be used with complete confidence, they must be checked out in final product form. Even a percentage point or two improvement in yield more than pays for such a program. But the starting ceramic powder is only one part of a very complex and interrelated manufacturing process.

Manufacturing MLCs

There are two primary approaches to the manufacture of an MLC capacitor. In the first, known as the *dry process,* a ceramic tape is cast on a moving belt using a doctor blade, dried, and recovered on a take-up reel. Electrode patterns are screened-on and the electroded tapes are stacked and laminated to yield an MLC "green cake" that contains a number of potential capacitors. Alignment of the successive layers during lamination is absolutely critical. Cutting separates the individual units, much as is done with semiconductor wafers, and firing creates the final rigid body. One of the principal advantages of this process is the ability to inspect the individual ceramic films prior to lamination.

The other process, pioneered in the 1950s by Sprague Electric, is the *wet process.* It is becoming increasingly important, primarily due to the ability to use thinner dielectric layers and therefore obtain greater volumetric efficiency than is possible with the tape process. While there are several different approaches, we will only describe the *curtain* or *waterfall* process, one of the most common in use today. In this procedure, a liquid paint is made in which the formulated ceramic powders are blended with the necessary solvents, binders, surfactants, and sometimes plasticizers. (This is essentially the same way house paint is prepared.) The ceramic paint is then continuously recirculated over a weir or dam to create a waterfall. A carrier, such a piece of 8" by 11" heavy cardboard, is passed rapidly under the waterfall and a thin layer of the ceramic-containing paint is deposited on it. The thickness of the final dielectric layer is controlled by the viscosity of the paint, the rate of flow of the paint over the weir, and the speed at which the substrate passes under the waterfall. The coated substrate is then dried and moved into a screener, where a liquid metallic paint, prepared in a manner similar to the ceramic paint, is screened-on and dried to form an electrode pattern. This electrode can be of expensive metals such as platinum and gold or, increasingly today, of a much less expensive silver-palladium composition in which the silver content is 70% or more. Work is also underway on the use of inexpensive metals such as nickel, but this is still a complex and difficult procedure.

This process is repeated continuously until the required capacitance is reached. The deposited metallic patterns are alternately offset to create the opposing capacitor plates once the individual chips are terminated. The resulting "green cake" can contain up to several thousand potential capacitors. These must be separated from each other, usually by either blade or saw cutting, and removed from the carrier prior to firing. As with the dry process, alignment is critically important during build-up.

To make the final capacitor, the separated chips go through a two-step firing process. In the first, the solvents, binders and other materials that have held the ceramic and metal particles together during the prior processes must be carefully burned-off so that no carbonaceous residue remains. This generally takes place around 500° C and can take several days to accomplish. Damage-free burn-off and uniform shrinkage require careful design of the ceramic and electrode paint systems

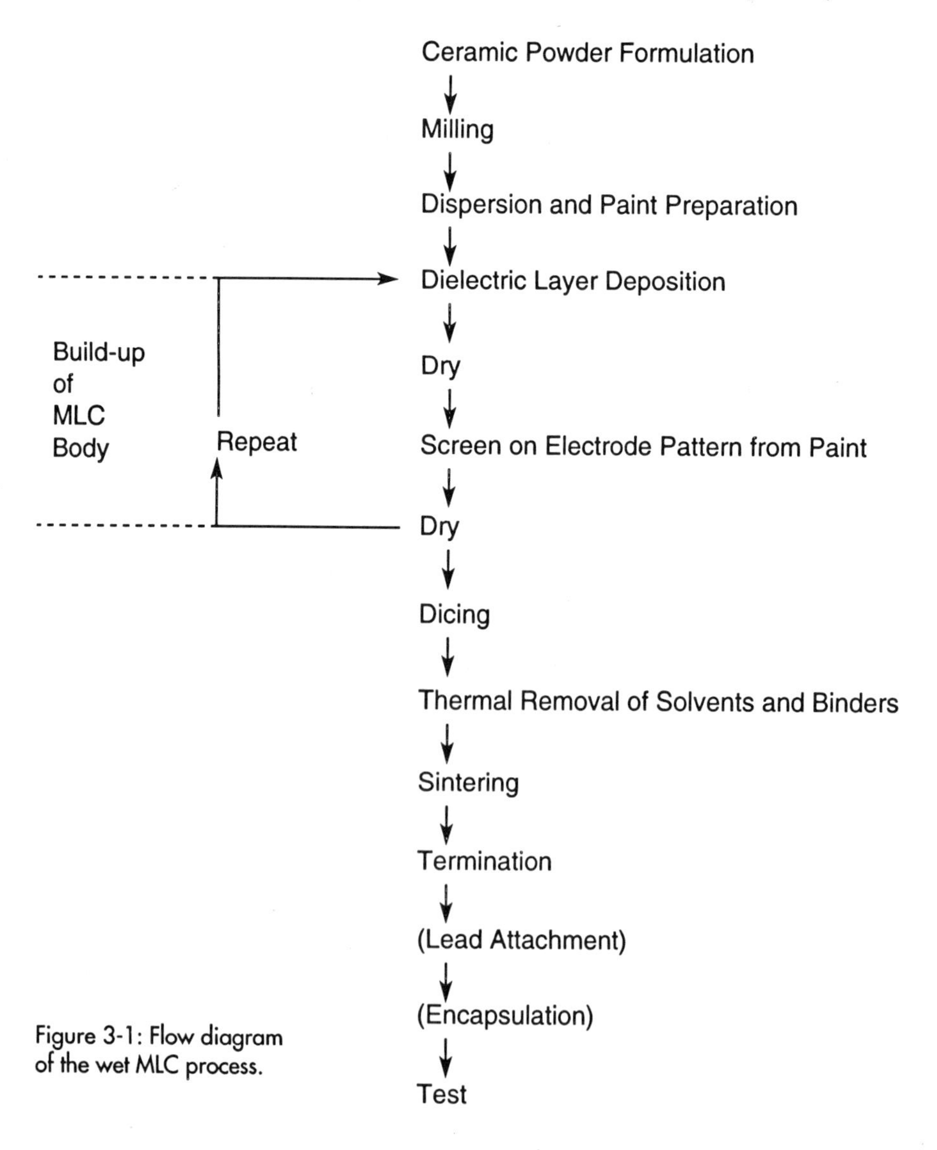

Figure 3-1: Flow diagram of the wet MLC process.

and tight control of the firing cycle. The second firing densifies the capacitor body and makes it into the same kind of rigid, rock-hard structure one finds in fired pottery. This can also take up to several days and, in the case of high silver content electrodes, occurs around 1100° C. To prevent delamination or cracking, this cycle must also be carefully controlled. Finishing involves termination of the sides of each chip with a metal such as silver (to connect the fired internal electrodes), lead attachment (unless one is making a leadless surface mount device, abbreviated SMD), encapsulation (if required), and test. The finishing process is essentially the same whether one uses a dry or wet process to manufacture the individual capacitor chips. Figure 3-1 is a flow diagram for the wet process we have just described, while Figure 3-2 is a conceptual diagram of the build-up section.

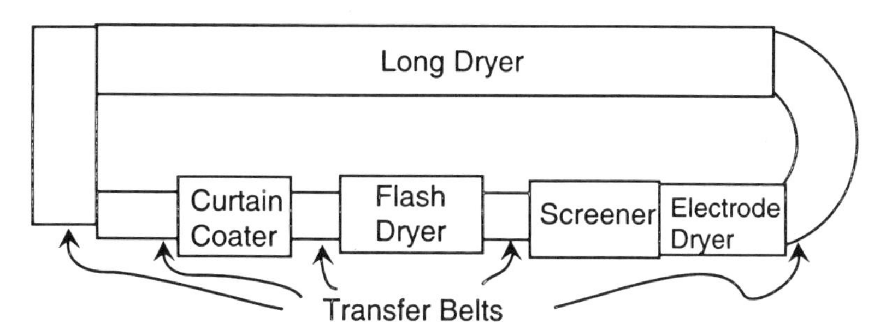

Figure 3-2: Conceptual layout of a waterfall or curtain "wet" build-up machine.

Ceramic Variables and Quality

Unless they are working with glazes, potters have many less variables to deal with than the electronic ceramicist. This is due to the fact that fine clays, such as kaolin, are so easily miscible and dispersible in water, eliminating the need for surfactants and plasticizers. In addition, the artisan is not dealing with the difficult problem of cofiring ceramic and metallic systems. What about formulations? Cathy only worked with three basic variations, while in electronic ceramics there is an almost infinite variety. However, in the capacitor field there are three primary EIA (Electronic Industries Association) specifications that interrelate the dielectric constant of the ceramic with its variation in capacitance over a specified temperature range (TCC). These are called COG (or NPO), with a K less than 100 and a flat TCC (less than 30 parts per million, abbreviated PPM) over the military range of -55° C to +125° C; X7R (or BX), with a TCC of + or -15% over the same temperature range and a dielectric constant generally below 3500;

and Z5U, with Ks up to 10,000 and a TCC of+22 to -56% from +10° C to +85° C. To understand the types of material and process controls that must be exercised in the MLC field, it is useful to take a closer look at a typical X7R formulation.

The major constituent is usually barium titanate, which makes up 85% to 95% of the formulation. However, other ingredients must be added to meet the required specification, improve performance, and allow use of lower cost electrode systems. For example, addition of the appropriate oxides of niobium (Nb) or neodymium (Nd) serves several purposes. Such "donor" dopants (Nb+5 for Ti^{+4} or Nd^{+3} for Ba^{+2}) inhibit grain growth during sintering and cause the presence of two phases in the fired ceramic. Each crystallite in such a "shell-core" structure consists of a central core of nearly pure $BaTiO_3$ surrounded by a shell of doped barium titanate. The two phases have different Curie points, leading to a flattening of the TCC over the temperature range. Addition of glass fluxes containing materials such as cadmium (Cd) or bismuth (Bi) lowers the sintering temperature and allows cofiring with high silver content electrode systems. Such fluxes also tend to lower the dielectric constant. However, use of high purity fine ceramic powders, tight control of particle size, and distribution as well as the ratio of barium and titanium creates sintered bodies with densities close to theoretical and Ks over 3000. Detailed theories have been developed to scientifically explain all these phenomena.

So what? Why all this detail and how does it relate to the relative competitive positions of U.S. and Japanese companies in a business such as ceramic capacitors?

People marvel at the complexity of the processes used to manufacture different types of semiconductor devices and integrated circuits. Yet in electronic ceramics we have identified an even more dizzying array of things that must be understood and controlled—composition, purity, particle size and distribution of both ceramic and metal starting materials; the role of additives and impact of impurities and defects; formation of ceramic and metallic paints and their compatibility during subsequent processing; accurate alignment of ceramic and metal layers during build-up; deposited layer thickness control; damage-free removal of solvents and binders during firing; shrinkage and densification during thermal cycles; dicing; termination; and the collected effect of all of these on product performance, yield, and reliability. Only by having some understanding of the impact of these many variables is it possible to evaluate how different approaches to a materials processing business such as MLCs can lead to success or failure.

Other Devices, Other Processes

Nor is this issue important only in semiconductors and electronic ceramics. Control of the properties of the starting materials and of the manufacturing processes is just as important in the other capacitor families, although the enormous complexity and variations that exist in ceramics is less prevalent. Several examples should make this clear. The aluminum electrolytic capacitor is generally constructed on high speed, automatic winding equipment and is a complex construction made-up of sequential layers of high purity aluminum foil, on which the insulating layer or dielectric has been formed by electrochemical oxidation ("anodization"), spacers such as paper, and a counterelectrode which is also made of aluminum foil. The "section" formed in this manner must also be impregnated with a conducting liquid or gel, called the electrolyte, which serves to make electrical connection between cathode foil and the surface of the alumina (Al_2O_3) dielectric that has been formed on the anode foil.

As I will show later in this book, the purity of the starting anode foil in an aluminum electrolytic is critical to the long term reliability of the device. The ability to increase the surface roughness or etch ratio of the foil is also an important competitive factor as far as volumetric efficiency is concerned. As discussed earlier, the capacitance or energy storage capability of a capacitor is directly proportional to the area of the dielectric film and, in the case of the electrolytic, this area can be dramatically increased by etching the surface of the aluminum foil prior to anodization. Obtaining a high degree of surface roughness is controlled by a variety of different and extremely complex processes. In addition, the nature of the electrolyte used in the finished device directly affects performance, especially as far as temperature coefficient of capacitance is concerned.

The construction of a solid tantalum capacitor is quite different, although the principles are the same. In the case of the solid, the anode is a high surface area pellet of sintered tantalum powder on which a dielectric film of tantalum pentoxide (Ta_2O_5) is anodically formed. As discussed in the previous chapter, the tantalum oxide insulator has the advantage of a considerably higher dielectric constant (26 versus 8) than alumina. In addition, the cathode is a solid material formed when the oxidized pellet is impregnated with manganous nitrate which is then transformed to semiconducting manganese dioxide (MnO_2) during a subsequent heating step. To make this cathode solderable for later lead attachment, the MnO_2 layer must be sequentially coated with graphite and then silver. The details of this complex sequence of process steps are important trade secrets to each competitor. As with the aluminum capacitor, the purity of the starting powder in a solid tantalum is equally important, as is the effect of particle size and distribution of the metallic powder on how much of the surface area of the sintered pellet actually contributes to the capacitance of the device.

The construction of a film capacitor is similar to the aluminum electrolytic in that the basic element is a rolled section involving alternating layers of electrode foils and paper or film as the dielectric, and is constructed on an automatic winder. However, there is obviously no equivalent to the anodization step in an electrolytic and most manufacturers purchase their dielectric films from outside vendors. As you might guess, the performance and reliability of the finished capacitor is heavily dependent on the perfection of these films as well as the processing steps that follow.

Of all passive components, materials technology is most important in the capacitor business, regardless of the dielectric system employed. This is because the dielectric is the key element in the device from every aspect. While capacitance increases with area, it is also inversely proportional to the thickness of the insulator. Therefore, for maximum volumetric efficiency, in addition to large area the dielectric layer should be as thin as possible without causing dielectric breakdown under the rated voltage of the capacitor. This holds true whether the dielectric is part of a wound section, as with an aluminum electrolytic or film capacitor; formed on a sintered body, as with a tantalum; or stacked, as with an MLC. This can only be accomplished if the starting materials are of high purity and there is absolute control over the subsequent manufacturing steps.

Although there are exceptions, with most other types of passive components control of the purity of the starting material is of somewhat less importance. For example, most inductive elements are made by winding wire around a core ferrite material. Here the cost of the high labor content involved is one of the most important factors in manufacturing. This is why such work is usually done in low labor cost countries such as Mexico, Taiwan, or Hong Kong. While connectors are more complex, the enabling technologies relate to the ability to form intricate metal stampings, shapes, and assemblies, and to expertise in the sciences of plating and of molding. An important exception to this is the currently small—about $100,000,000 in annual sales in the U.S.—but increasingly important filtered connector segment. Here inclusion into the connector assembly of ceramic capacitor or inductive-capacitive (LC) filter elements can involve materials and processing technology that is just as complex as with an MLC.

While a carbon composition resistor is extremely simple from a materials standpoint, things get more complex with precision wire-wound and metal film resistors. In the case of thick film discrete resistors, resistor networks, and resistor-capacitor networks, the processing becomes more complicated. Using a substrate such as alumina, the manufacturing process may involve sequential deposition and firing of glazes, metal interconnects, resistive materials, dielectrics, and electrodes. In certain precision applications, laser trimming may also be required to bring resistors to the final desired values. Such thick film technology can be very complex and capital intensive.

The addition of transistors and/or integrated circuits to such substrates can be used to make hybrid integrated circuits or, if the simple substrate is replaced by one in which multilayer ceramic capacitors and interconnects have been incorporated, extremely complex electronic circuits and systems can be created. Before its exit from the ceramics business, Sprague Electric had a pilot capability in such an approach which they trademarked "Multilythics." Figure 3-3 is a conceptual drawing of a complex circuit formed using this technology. In this case, all the capabilities necessary in both thick film processing and in multilayer ceramics are required. This expertise was sold to LC Thomson of France several years ago.

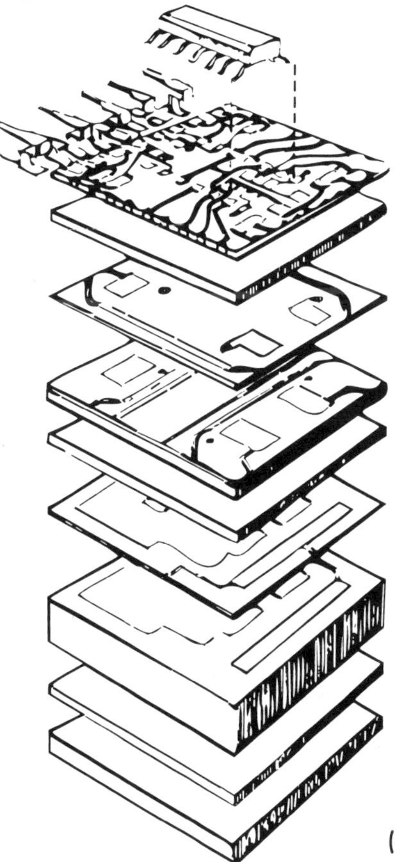

Figure 3-3: Complex functional circuit (courtesy of Sprague Electric).

Having already discussed electronic ceramics technology in some detail, it seems unnecessary to give a more complete review of the processes used to manufacture these other component families. However, at this point it seems clear that in semiconductors, capacitors, and many other component businesses, those companies which succeed should be those which have the greatest understanding, control, and competitive advantage in the materials and processes they employ. We will discuss and greatly expand upon this point as we compare how the U.S. and Japan approach such industries in the remainder of this book.

Success and Failure at Sprague Electric: Early Sources

I can say with certainty that much of Sprague Electric's early corporate success was due directly to its devotion to materials-related research and development and to having a superior capability in materials science. This was due partly to the fact that at that time technology companies were founded, run, and staffed primarily by scientists and engineers. It was also partly due to the fact that my father was particularly adept at hiring and keeping some of the very best technical talent in the infant industry. Perhaps most importantly, he was an engineer by training and personally dedicated to technological leadership in whatever product areas his company decided to pursue.

If the reader who is unfamiliar with Sprague Electric and its history is somewhat perplexed at this point, I am not surprised. So far, I seem to have described two completely different companies. On one hand, we have a corporation that started from scratch in the 1920s, was a worldwide technology and business leader in the capacitor field for many years, and that was also able to develop a successful niche business in the very difficult semiconductor market. On the other hand, in recent years this same company sold its entire semiconductor capability to the Japanese and exited the only real growth part of the capacitor industry, multi-layer ceramics or MLCs. This apparent dichotomy is due to the fact that what I have really portrayed is different times in the life cycle of a corporation. The reasons for what happened to Sprague will become clear as this book progresses. Most of those reasons parallel what has happened to all too many other U.S. electronic component companies.

At this point, suffice it to say that Sprague made at least two fundamental mistakes in the late 1960s and 1970s. First, it did not have sufficient resources—either in competent management or finance—to succeed in all the business enterprises it attempted. Unquestionably, management's fixation with success in semiconductors detracted from the core capacitor businesses. Secondly, it didn't recognize until almost too late that Japanese companies would eventually develop the capability to be formidable competitors in most of Sprague's different capacitor families. The initial Japanese offerings were primarily miniature aluminum electrolytics for consumer electronics and automotive radio applications. The early performance and reliability of these devices were suspect, and Sprague's initial response was to try and prevent dumping of such components into the U.S. marketplace at unrealistically low prices. Unfortunately, this effort met with little success.

Since materials science and technology are such critical elements to success or failure in the electronic components business, it now seems appropriate to turn to a more detailed analysis of how technology and innovation are practiced in the U.S. and in Japan. That will be the subject of the next two chapters.

CHAPTER 4

Sources of Competitive Advantage

Some American high technology systems were given the ultimate field testing during 1991's Operation Desert Storm. The images of that period are unforgettable: "smart bombs" swooping down airshafts, Patriot missiles knocking down Iraqi Scuds, and stealth bombers operating with impunity over Iraqi airspace. Highly criticized systems, such as the Bradley fighting vehicle and MIAI Abrams tank, proved far superior to anything they faced, including the best Soviet military hardware. Not all the success was in hardware; the all-volunteer U.S. armed forces operated with precision and extraordinary efficiency. It was perhaps the most impressive total performance ever by the American military.

Whether 1991's Gulf War was worth the cost or achieved its objectives is beyond the scope of this book. What does concern us is what it told us about American capabilities. First, it showed that, given proper incentives and financial support, the United States can design and manufacture the most complex electronic systems ever created. Secondly, the war proved that properly trained and motivated Americans (i.e., the troops) can maintain and use such systems under the most difficult operating conditions. This means that, given the proper education and training, our people are as capable as any.

But, if these observations are correct, what is there about our industrial process that causes us to have such difficulty in competing in non-military electronic businesses? Where does the root of this problem lie?

While some readers might disagree, I believe we, as a nation, must all first understand and accept that the U.S. does have a basic competitiveness problem. With this being the case, let's examine the comparative sources of competitive advantage in both the United States and Japan in an attempt to answer such questions.

In trying to answer such questions, I obviously will be drawing upon my own experiences in the U.S. and Japanese electronics industries. But I don't claim to have a monopoly on expertise or insight. In discussing the relative strengths of the United States and Japan, I will frequently refer to the books listed below. They were chosen primarily as the result of reviews I had read or because a particular text had been recommended to me by one or more business associates, and have been particularly helpful in supplying an overall perspective to complement my own background:

- *Competitive Edge: The Semiconductor Industry in the U.S. and Japan*, edited by Daniel I. Okimoto, Takuo Sugano, and Franklin B. Weinstein. Published by Stanford University Press, Stanford, California, 1984.
- *Microcosm: The Quantum Revolution in Economics and Technology*, by George Gilder. Published by Simon and Schuster, New York, 1989.
- *Kaisha: The Japanese Corporation*, by James C. Abegglen and George Stalk, Jr. Published by Basic Books, Inc., New York, 1985.
- *Kaizen: The Key to Japan's Competitive Success*, by Masaaki Imai. Published by Random House Business Division, New York, 1986. Several of my Japanese (and American) contacts feel this book is the best single description of the reasons for Japan's competitive advantages over the United States.
- *Innovation: The Attacker's Advantage*, by Richard Foster. Published by Summit Books, New York, 1986.
- *Made In America: Regaining the Productive Edge*, by Michael L. Dertouzos, Richard K. Lester, Robert M. Solow, and the MIT Commission on Industrial Productivity. Published by Harper Perennial, New York, 1990.
- *The Competitive Advantage of Nations*, by Michael E. Porter. Published by The Free Press, New York, 1990.
- *Japan in the Global Community*, edited by Yasusuke Murakami and Yutaka Kosai. Published by the University of Tokyo Press, Tokyo, 1986.

While I recommend these titles—as the following discussion shows—I don't always agree with their opinions and conclusions. With these titles as source material, let's see how the United States and Japan stack up against each other.

Advantages: Japan

As you might expect, Japan has several advantages, especially when it comes to human and social resources. Let's examine them.

Teamwork

One question that is often raised is whether or not there are ethnic sources for the many advantages seen in the current Japanese economy. Murakami and Kosai address this in *Japan in the Global Community*. They offer some interesting perspectives on this question, and point out that there are important cultural differences. This book claims that there are four historical and geographic factors which have made major contributions to the Japanese character.

The first was the emergence of a society based on rice agriculture; this developed a strong reliance on cooperative group activity. This observation has also been made by several of my own contacts in Japan. The reason given is that the large amount of cooperative labor required in rice farming has led to a society that is inclined to operate more successfully through group endeavor than through strong individualism.

Secondly, because of their relative isolation as an island nation, the Japanese were initially forced to learn much on their own and, as a result, they had to develop a strong, shared information network. This tendency to openly share information and the desire on the part of each individual to be an important contributor to the group led to an emphasis on attention to detail and group decision making.

Third, the emphasis on diligence, hard work, and frugality, as well as the importance of mass education, came from both the influence of Confucianism, which followed the decline of Buddhism as the ruling faith in Japan that began in the sixteenth and seventeenth centuries, and the fact that the nation is almost devoid of natural resources. Only through hard work and diligence is it possible to succeed.

The fourth stated factor is, to me, both the hardest to logically understand and also the most interesting. *Japan in the Global Community* claims that the Japanese developed as "skilled learners" because they were both isolated from unwanted outside influences and yet still close enough to the Asian continent to exchange information where they felt it was necessary. In other words, they were able to learn and adapt only what they wished, without "intellectual and political domination by foreign powers." This report, which was published in 1986 as the result of a series of roundtable discussions organized over two years by the Ministry of International Trade and Industry (MITI), gives an important glimpse at how the Japanese view the outside world: take only what is in our best interests. This still does not really answer why they are such good learners. I believe this results from both a very strong emphasis on education and, because of the emphasis on group behavior, an almost complete absence of intellectual NIH (not invented here) syndrome.

Japan in the Global Community also claims that the group effort required by rice agriculture has led to a society where strong leadership is not very important. I find this observation hard to accept. Whether at the CEO level, or elsewhere in the organization, most Japanese managers I have met and interfaced with have been very strong leaders and individuals.

Education

Most authors agree this is another area of Japanese advantage, often to a major degree, at the primary and secondary levels. For example, the authors of *Competitive Edge* say that the Japanese workforce is probably the best educated in the world. The authors of *Made in America* specifically cite weaknesses in primary and secondary education in the United States as hampering American competitiveness. Porter, in *The Competitive Advantage of Nations*, also cites the weaknesses in the American education as one part of a general neglect of human resources in the United States. However, Japan is still outspent on education by the United States. At the time this book was written, the United States was spending 6.8% of its gross national product on education versus 5% for Japan (and 4.5% for Germany). Clearly, a lack of funding is not the reason for the failures of American education.

Labor

This another area where most authors agree Japan has an advantage. In *Made in America*, the authors cite the cooperation between labor and management. Such ambitious programs as just in time (JIT) manufacturing and total quality control (TQC) can work only if the labor force gives them support.

Japanese labor unions are frequently described as "cooperative" (as in *Competitive Edge*) but, from an American or European perspective, perhaps "docile" would be more accurate. Regardless of the description used, the fact remains that Japanese companies generally have smoother relations with the unions representing their employees than do American firms. Another difference is that Japanese labor unions have usually tended to welcome factory automation and work simplification instead of fighting it.

Business-Government Relations

This is an area where most authors also agree Japan has an advantage, even though they don't agree on exactly "what" that advantage is. Perhaps the most extreme interpretation of this advantage is the concept of "Japan, Inc.," which postulates that Japanese firms are simply executing a clever government plan to achieve world domination in selected industries. This image shows up frequently in the American media and political debates.

However, the concept of "Japan, Inc." does not find support among the authors we've cited so far. In *The Competitive Advantage of Nations*, Porter dismisses the notion by pointing out Japanese companies are intensely competitive with each other, particularly in their domestic markets. He says the role of the Japanese government is more in "signaling" which industries it feels are most important rather than outright direction and control of what those companies do. *Made in America* likewise finds no convincing evidence for some sort of grand government/industry conspiracy to dominate world markets. Moreover, Japanese companies need no governmental prod to grow as fast as possible and expand market share. Abbegglen and Stalk write in *Kaisha* how Japanese corporations already have an overwhelming growth bias and drive for ever-increasing market share. Any Japanese government action would seem to just reinforce strong trends that already exist.

Much more difficult to measure or assess accurately is the attitude and policies of the Japanese government toward business as compared to the American attitude. This task is complicated by the fact that, for example, American semiconductor firms must interact with local, state, and national governments, and the attitudes and policies of these different levels of government are not always consistent or even compatible with each other. However, *Made in America* bluntly cites a generally antagonistic relationship between business and government in the United States as a major reason for a lack of American competitiveness. Certain Japanese government policies, like those giving preferred tax treatment to interest on personal savings and capital gains, clearly encourage investment. Japanese companies can also cooperate in ways that would probably lead to anti-trust action in the United States. On the issue of industrial policy, it is clear that Japan has one, although not to the extent that proponents of the "Japan, Inc." theory would like to believe. However, the Japanese government does take a hand in targeting certain industries through joint research and development coordinated by MITI. Whether the United States needs or should have an industrial policy is the subject of much political and economic debate. Some would argue that the United States government has long had an industrial policy—one directed toward producing the finest military hardware in the world. Porter, in *The Competitive Advantage of Nations,* says, however, that defense-related research and development is more of a distraction than an advantage in the industrial and consumer areas. *Competitive Edge* also notes there is currently limited industrial spin-off from defense-related research and development, which is in contrast to the early days of the U.S. semiconductor and computer industries.

Inexpensive and Available Capital

Competitive Edge cites the ready availability of capital at a low cost through financial institutions as a clear Japanese advantage. Abegglen and Stalk expand on that point in *Kaisha,* noting that Japanese companies are able to make investments ahead of demand instead of after. This lets Japanese companies get further along the learning curve and reduce costs faster. The cheaper costs of capital let Japanese companies assume greater fixed cost burdens both in equipment and human resources. *Made in America* also cites the cost of capital as a problem for U.S. competitiveness, although the authors see it as part of a matrix of difficulties rather than as the determining element.

Capital is not only cheaper and more available in Japan, but it is also different. In Japan, bank loans account for the major funding of most semiconductor companies. In the United States, most funding is through equity such as stock or venture capital. A bank loan is inherently a longer term investment than a share of stock that can be bought today and sold tomorrow. This difference in how companies are capitalized affects how they operate. Japanese companies can be (and, from outward appearances, are) managed with an eye toward long term results—since that is what is of main interest to bankers making long term loans—instead of quarterly numbers. American companies, on the other hand, typically face demands by equity owners for quick returns on investments and favorable quarterly results. As *Made in America* points out, many U.S. companies operate with short horizons. This is often manifested by ignoring commodity products with high volume potential (like DRAM chips) in favor of products that offer a higher initial return on investment, and ignoring markets where success may be a few years down the road. Since one of the quick and easy ways to increase quarterly profits is to cut variable costs, American companies tend to slice investments in training, capital goods, and research and development when times get difficult. American managers are often criticized for making decisions that bring short term benefits at the expense of long term gains. However, they are often just doing what the ultimate company owners—stockholders and others with an equity interest in the company—demand they do. In contrast, Japanese bankers and most equity owners are willing to be patient for bigger payoffs that may be years in the future.

In addition to tax policies that favor savings, the Japanese have other strong incentives to save. These include high housing costs that make home ownership difficult (if not impossible) for most consumers, a lack of consumer credit, and—because of less attractive pension, social security, and health care systems—a need to save personally for retirement. Not only are Japanese "social security" (their equivalents to American programs) and company pension benefits lower than in the U.S., there is a five year gap between mandatory retirement at the age of 60 and age 65 when social security becomes fully available without actuarial reduction. This means that citizens must both save and find part-time, sometimes menial, work to cover this period. In effect, the lack of a comprehensive social "safety net" such as that found in the United States compels the Japanese to save more for their retirement and to cover emergencies.

A final consideration is that Japanese semiconductor firms are much larger and more diversified than their U.S. counterparts. For example, Mitsubishi makes semiconductors in addition to steel, autos, ships, and other items. Toshiba is involved in a wide range of consumer electronics manufacturing as well as semiconductors. Size is not always an advantage for a company, but it usually makes accumulation of large amounts of capital on favorable terms easier.

Ability to Exploit Technological Innovation

If American managers can be somewhat excused for their fixation on short term results, the same charitable view cannot be taken of their failure to exploit new technological developments and commercialize them.

Foster provides depressing examples of this in *Innovation: The Attacker's Advantage.* He points out that American vacuum tube manufacturers, such as General Electric, Westinghouse, Sylvania, RCA, and Raytheon, were slow to respond to the challenge posed by the transistor and were worried about its impact on their profitable tube businesses. Instead of recognizing that the future of electronics was solid-state, they hesitated and let aggressive startups like Fairchild, National, and Intel dominate the new industry. (However, Motorola was an established company that recognized the potential of the transistor and today is a major world competitor in transistors and ICs.) There are many other cases which Foster does not mention that I can recall. For example, Digital Equipment Corporation (DEC) was slow to enter the personal computer market because they were afraid of hurting sales of their minicomputers. Instead, personal computer companies like Apple and Compaq hurt sales of DEC minicomputers, and DEC is now struggling.

American managers also fail to exploit new ideas. The teachings of Deming and Juran have recently become close to holy writ in many corporations, and the concepts they pioneered—such as total quality control and just in time manufacturing—are the latest corporate buzzwords. However, there is nothing new in these ideas, as Deming and Juran have been articulating them since the early 1950s. At that time, however, American managers were not listening to them—only Japanese managers were.

Fiercely Competitive Home Markets

Japanese companies may be rough on American companies, but they are rougher on each other in the Japanese domestic market. This point is made by Abegglen and Stalk in *Kaisha* and Porter in *The Competitive Advantage of Nations*, and is clear to any visitor to Japan who has observed the domestic competition found in such areas as consumer electronics and automobiles. In the United States, we do not get a true idea of how fiercely competitive the Japanese domestic market is since not every item of consumer electronics or make of automobile made in Japan is available in the United States. There are some large Japanese companies (such as Daihatsu in automobiles) which only sell to their domestic market and do not export. However, they do fight for the Japanese consumer's cash and increase the pressures on companies that do export. The result is that the Japanese companies in the American market have been battle-hardened in Japan and tend to be "the best of the best."

Kaizen: The Spirit of Improvement

In his book *Kaizen*, Masaaki Imai defines "kaizen" as the continuous drive to make incremental improvements in all aspects of personal, social, and business life. This was very much in evidence in many of the companies I dealt with in Japan, including Sprague Electric's former Kyoto-based joint venture, Nichicon-Sprague. During the plant tours that were an important part of every visit I made to N-S, there were always some bright young engineers proudly demonstrating the new pieces of equipment that had been added to the manufacturing floor since my last visit. The new equipment was usually not part of some major project; more often, the new items were relatively simple handlers that eliminated hand labor operations and improved both quality and productivity. Kaizen is very much part of the entire quality thrust in Japan, and is one reason why the United States has had so much trouble trying to catch Japan in quality—the U.S. is trying to catch a moving target. Porter, in *The Competitive Advantage of Nations*, also notes the importance of kaizen.

Process Development and Manufacturing Ability

Here is an area where most observers agree Japan has an overwhelming advantage over the America. The authors of *Competitive Edge* say that Japanese firms concentrate their engineering and financial resources in areas of process innovation and manufacturing technology, with continuing investment in capital equipment giving improvement in both productivity and quality. Foster in *Innovation* says that U.S. companies focus more on new product development while their Japanese counterparts concentrate on process development and manufacturing technology. This becomes a powerful competitive advantage for the Japanese as a business matures. Other books that cite a Japanese manufacturing advantage are *Made in America* and *The Competitive Advantage of Nations*. In the latter book, Porter says neglect of human resources is a major reason for U.S. manufacturing shortcomings.

An Emphasis of Manufacturing

In recent years, there has been much written about the United States becoming a "service economy" rather than one based on manufacturing. Some have said that the loss of U.S. manufacturing jobs doesn't really matter, as new service jobs will replace them. The Japanese (and the Germans too, for that matter) do not subscribe to this view. Manufacturing is viewed in Japan as the key source of a nation's wealth, and the Japanese emphasize manufacturing for that reason.

Made in America argues that manufacturing is necessary to create goods for export because of America's seemingly insatiable desire for imported goods (services are obviously much more difficult to export). In the October, 1991 issue of *Technology Review*, Bennett Harrison points out that manufacturing has a much greater leverage than services in creating jobs in the support infrastructure. In addition, manufacturing jobs often have higher wages than service jobs, and there just aren't that many service jobs available in an economy without a strong manufacturing base. (Fewer manufacturing jobs mean a reduced need for cost accountants and industrial engineers, for example.) The manufacturing floor is where a lot of innovation and skills development takes place. Harrison also notes that the advantages of being "first to market" generally come through manufacturing and not services.

At the time this book was written, there were signs that the American people (and politicians) were finally beginning to realize the necessity of revitalizing our manufacturing sector. However, it will take extraordinary efforts before the United States can begin to approach Japanese manufacturing capabilities.

Clustering

In *The Competitive Advantage of Nations*, Porter uses a concept he terms "clustering." By this, he means the presence of related and supporting industries in close proximity to a major industries. These related and supporting industries include suppliers, machine shops, and various material and service subcontractors. This is sometimes found in the United States, as in such areas like Route 128 in Massachusetts and Silicon Valley in California, but not to the extent seen in Japan. This is because Japanese companies actively try to bring about such clustering of the necessary support infrastructure around them.

As Japanese companies set up manufacturing facilities overseas, they are following this same approach with their local suppliers. The importance of clustering will become clear later when we examine the erosion of the infrastructure supporting the U.S. semiconductor industry and the competitive advantages of powerful Japanese industrial groups known as *keiretsu.*

Advantages: United States

While Japan has numerous advantages, so does the United States.

Basic Science

As we saw back in Chapter 2, Japan has made few significant contributions to basic electronics knowledge. The authors of *Competitive Edge* as well as Foster in *Innovation* acknowledge America's advantage in basic science. Imai in *Kaizen* seems to feel that the U.S. advantage in this area is not as important as it might seem. I feel Imai's conclusion reflects in part that he does not have an engineering background and cannot fully appreciate the importance of basic research.

It also might express a feeling common among many Japanese that the necessary technology will always be available, whether from domestic or foreign sources.

Product Development

This concept is closely related to our previous one. Again, the historical record discussed in Chapter 2 shows few breakthrough items (like the microprocessor) from Japan. Perhaps the most optimistic assessment of this advantage is by Gilder in *Microcosm*. In that book, he argues that this advantage will become even more important in the future as the need for application-specific integrated circuits (ASICs) increases, and ASICs play directly to the U.S. strength in product development. However, I dispute Gilder's assessment. Obviously, not everyone in the electronics industry can be an IC designer, and manufacturing commodity ICs generates more total jobs than manufacturing ASICs. Moreover, Gilder ignores the fact that "U.S. engineers" are less and less U.S. nationals. Our engineering schools are filled with students from Asia. Many of them return to their country of origin after spending some time with a U.S. firm. Nor is product creativity exclusively an American trait, as Gilder implies. Take the case of Masatoshi Shima, who we met back in Chapter 2. He was a key figure at Busicom when Intel developed the first microprocessor for them. He joined Intel in the U.S. in 1972, and then moved on to Zilog, where he designed the famed Z80 microprocessor. He rejoined Intel and returned to Japan in 1979 to set up Intel's Japanese design center. I feel there are many more Masatoshi Shimas out there.

Total Research and Development Expenditures

The authors of *Competitive Edge* cite a single fact that many people will probably find surprising: total expenditures for research and development in the United States are almost three times greater than in Japan. You might think this difference is due to defense needs, but only about one-third of the U.S. amount is (at the time this book was written in 1992) funded by the defense budget. This means that the United States outspends Japan by almost two to one on non-defense research and development work. It is interesting to note that Masatoshi Shima's development work was done for two American companies—Intel and Zilog—instead of Japanese firms. This clear American advantage would be even more important if U.S. companies could better commercialize the fruits of their research and development funding.

University System

The authors of *Competitive Edge* also give high marks to the U.S. scientific university system, as do the many foreign students who enroll each year. Gilder makes the same point in *Microcosm*. Given the excellence of its primary and secondary educational system, the Japanese university system is, as a whole, surprisingly mediocre. While the universities at Tokyo and Kyoto are world-class, many Japanese college students spend their university time in rote memorization of material rather than creative, original scholarship. This stultifying university environment may be one reason the Japanese are weak in basic scientific research.

Venture Capital System

Perhaps the greatest impediment to creative Japanese nationals like Masatoshi Shima is the lack of viable alternatives if one of their ideas is turned down by their employer—they are essentially stuck. As we'll see shortly, mobility between companies in Japan is very limited. And the venture capital industry is very much in its infancy in Japan. By contrast, getting an idea rejected by an employer has been the starting point for a lot of American electronics companies. Disgruntled engineers who really believe in an idea can usually find a venture capitalist somewhere who will listen to them. Companies such as Apple Computer, National Semiconductor, Intel, Linear Technology, Compaq Computer, Silicon Graphics, and many others have been launched by unhappy engineers and venture capital funding .

There have been recent efforts to develop a venture capital industry in Japan. However, these received a severe setback in December, 1991 with the untimely death of MIT-educated Yaichi Ayukawa, then president of Tokyo-based Techno-Ventures Co., Ltd. It is unlikely that anything resembling the U.S. venture capital industry will be found in Japan in the near future.

Natural Resources

Here is another area where the United States has a major advantage over Japan. Perhaps no other nation has such a favorable array of natural resources at its disposal. These include both renewable resources, such as agriculture, and "one time" extractable resources such as petroleum and other minerals. By contrast, the Japanese have very few natural resources other than low-grade coal, and must export in order to pay for their import needs. Japan is vulnerable to any sort of unexpected disruption in key resources. For example, the loss of all oil supplies from the Middle East would cause extreme hardship for the United States, but at least the U.S. would have its own domestic sources to fall back on plus pipeline access to the oil fields of Canada and Mexico. For Japan, the loss of Middle East oil would be a catastrophe of almost biblical proportions since Japan imports all its oil via sea routes. The extreme Japanese dependence on imported resources means it is also dangerously vulnerable to any sort of naval war or other action that would restrict ocean shipping. This dependence of resources from the outside world helps explain the persistent—and, to many Americans, often baffling—sense of insecurity and fatalism often found in Japanese foreign policy, political life, and lower-level managers. Porter makes this observation in *The Competitive Advantage of Nations*, where he says pessimism and insecurity drive many Japanese managers. However, I have yet to meet a Japanese CEO who displays these traits!

Lifetime Employment: Realities and Drawbacks

Quite a few Americans feel the "lifetime employment" policies of Japanese companies are a key competitive advantage for those firms. Freed from having to worry about and protect their jobs, some arguments go, workers can concentrate on being productive and innovative. Several public figures have advocated that U.S. companies adopt similar policies. But just what does "lifetime employment" in Japan mean?

In trying to answer that question, I received input from a number of different sources, including the Electronics Industry Association of Japan (EIAJ). By law, lifetime employment is meant to apply to all full-time employees of large Japanese corporations. However, I never got adequate definitions of what "large" and "full-time" mean under the law. My research showed that most major Japanese firms have full-time employee staffing which they try to maintain at relatively stable levels. Any variations in staffing needs are absorbed by part-time employees or subcontractors, neither of which are covered by the lifetime employment law. Further, women traditionally quit their jobs when they get married (the social pressures to do so are enormous and compelling) and generally do not return to the work force except as part-time employees. Except at certain executive levels, the *mandatory* retirement age in Japan is 60. (Protection against age discrimination, as in the United States, is essentially unknown in Japan.) Subtracting out such workers exempt from the lifetime employment law, my best estimate is that slightly less than one-third of all Japanese workers actually have "lifetime employment," and those are almost exclusively males under 60 years of age.

Many American advocates of lifetime employment seem unaware the practice comes with some strings attached in Japan. The biggest one is a lack of mobility; for those covered by lifetime employment, their first employer is often their only employer. Most large Japanese corporations have what Murata Manufacturing Co., Ltd. refers to as their "freshman program." Murata takes in several hundreds of college graduates each year, with a large percentage being engineers. They are guaranteed employment at Murata for life (i.e., up to age 60) unless they choose to leave. The new hires are cross-trained throughout the company for a number of years before settling down in a specialized area.

The benefits to companies like Murata are clear—a stable number of well-educated, loyal employees. But suppose an employee becomes disenchanted with Murata and wants to work elesewhere? Movement between companies in Japan is rare, and usually occurs because the hiring company wishes to gain a special skill rather than because the employee is looking for a better deal. Hiring away employees of another company is viewed by most company CEOs as highly unethical behavior. I remember well the fury expressed by President Hirai of Nichicon when two of his engineers were hired away by an important competitor, Nippon Chemicon. His anger was directed at both the engineers, because they had violated what Hirai considered to be a sacred trust, and Chemicon, because they had broken an unwritten agreement of not hiring employees away from a competitor. (To be fair, this sort of unwritten, informal agreement has existed between major U.S. capacitor competitors for many years.) This lack of mobility is one reason why it is very hard for U.S. companies to hire professional and managerial employees in Japan unless such companies have been established there for years and are considered "Japanese," like IBM-Japan. As we've previously discussed, the infant Japanese venture capital industry makes starting one's own company not a real option. While mobility between companies is becoming easier due to skill shortages, from an American perspective lifetime employment in Japan sometimes seems more like indentured servitude. It's unlikely that any similar system could work in a diverse, individualistic American society; how many American employees would be willing to trade their freedom to move to another job in return for "lifetime employment"? And such a system would probably reduce the number of full-time employees corporations would hire. Lifetime employment would make personnel costs a fixed cost rather than a variable cost, and—as in Japan—would force companies to keep the total number of full-time employees to a minimum and use part-time employees and subcontractors to fill in any gaps in staffing needs.

Is lifetime employment a source of competitive advantage in Japan? Since we're interested in the competitive relationship between the United States and Japan, we'll ignore whether this practice is good or bad for the workers involved. Perhaps the fairest answer is that it is a two-edged sword. Japanese managers don't have to worry whether an entire design team will walk out the door to start their own company. Japanese employees don't decline transfers, relocations, or difficult assignments. Investments in human resources and training pay bigger dividends in Japan due to work force stability; too many American companies provide employee training only to see employees take their new skills to another company. However, lifetime employment clearly reduces innovation and technology diffusion between companies. Employees learn only one way—the Murata way, for example—and are required to completely support existing corporate policies and goals. The sort of questioning of the status quo that makes possible a breakthrough development is rare within a Japanese company. Since employees cannot freely move between companies, new ideas and different perspectives are often hard to come by. (We'll soon examine how Japanese companies must go to extraordinary lengths to gather technology information.) Indeed, would it have been possible to found a company like Apple Computer if the United States had a comparable lifetime employment system and lack of venture capital sources?

The analysis we've just done of the relative competitive advantages of the United States and Japan lead to the conclusion that the United States has the upper hand in basic research and innovation while the Japanese are superior at turning innovations into commercial products. In our next chapter, we'll look at how Japan "commercializes" innovations.

CHAPTER 5

From Basic Research to Actual Products

If Japan has not been historically a center of research and innovation, how have they obtained the technology and methods they have built their current prosperity on?

There are three ways to answer this question, and we will explore all three in this chapter. One is that the Japanese have obtained technology and methods from the outside through such methods as "gatekeeping," alliances with other companies, and licensing of technology. The second answer involves the way the research and development function of a corporation is viewed within a company and how it is managed. The third answer involves the commercialization of technology. All three answers point to the same conclusion: despite lacking strong basic research programs of their own, Japanese companies have succeeded in fields such as semiconductors through a disciplined methodology of accessing technology—wherever it is found—by creative methods and then adding value primarily in the areas of commercialization, process engineering, and manufacturing.

Gatekeeping

Anyone who has ever written an experimental Ph.D. thesis understands how important it is to understand all prior art. One of the greatest concerns of the candidate is to be sure that the work is truly new (and hope that no one scoops it prior to acceptance by the university). Back in the 1950s, when I was at Stanford, keeping up with the literature was not that difficult. I did so primarily through *Chemical Abstracts.* It is much more complicated today. In addition to books, journals, patent literature, a seemingly endless series of technical conferences (with proceedings often published well after each conference and sometimes only as abstracts), and unclassified government contract reports, one should also tap database systems (such as Dialog), keep track of university research, attend trade shows and seminars, look for license opportunities, and keep in continual contact with other experts in one's field around the world. Other useful sources include suppliers, customers, analysis of competitors' products, and, occasionally, even competitors themselves. This wealth of information must be systematically analyzed, prioritized, and incorporated into the firm's strategic and technology plans.

Despite language difficulties, Japanese corporations have been extraordinarily adept at collecting available information worldwide. We used to laugh at the number of Japanese engineers that attended technical meetings in the U.S. and tried to visit our factories, armed with their cameras and endless questions. We laugh no more.

There are a number of reasons for this enthusiastic Japanese drive for information. Some are cultural, such as their general thirst for knowledge and low level of NIH (not-invented-here) syndrome. Since they grow up in a society where information is extensively and openly exchanged, it seems perfectly natural to a Japanese engineer or scientist to tap all available sources of information. They are also culturally very thorough. Because of this, Japanese engineers and scientists seek all intelligence that is available on a subject of interest. Other reasons are purely practical. Japanese technologists recognize that most basic science, and much new product development, still takes place in the West, especially the United States. To access it, they are forced to seek out the many sources where it is available. Since there is little movement between corporations in Japan—and mergers, acquisitions and spin-offs seldom occur—individual corporations jealously guard their own trade secrets and proprietary information, and are forced to look outside of Japan for input. The openness of Western culture makes its technology relatively easy to access. This is also one of the reasons why Japanese companies have been aggressive at sending bright young engineers and scientists to the U.S. to study at our scientific universities.

Most studies on innovation detail the need for a "gatekeeping" function. This is typically one or more people in the technical organization who keep track of what is going on in fields of interest around the world. Sprague Electric always had several, and this meant that many of the rest of us did not need to read the literature as much as we should. The gatekeeper(s) kept us on our toes. For all the reasons given previously, my personal observation is that the average Japanese scientist and engineer is a better gatekeeper than his American counterpart. By this I mean that most Japanese technologists religiously follow the literature while, in the U.S., it is a much smaller percentage of the scientific population. It was always frustrating to visit Sprague's well-equipped research library and see how few of our scientists actually used the facility on a regular basis. It always seemed to be the same people who, incidentally, were also the best researchers. The Japanese competence in this important function is all the more extraordinary when one realizes how much of the scientific literature today is in English. On the other hand, due to inferior language skills, most U.S. scientists are at a severe disadvantage when a publication is in a foreign language, especially if it is Asian.

Licensing

Licensing is another important source of technology. It must be recognized that there are two forms: one of which relates to patents, and one to the detailed technology itself or know-how. Sometimes they go together and sometimes they don't. When a company has one or more basic patents in a field or a large portfolio—or hopefully both—they can make it impossible for another corporation to manufacture and sell in this field without a patent license agreement. Such can involve both an upfront payment and license fees paid over the period of time the patents are active. These fees can range from a few percentage points on sales to, on rare occasions, as much as 10%. As a result, companies with active patent filing and licensing programs can generate lucrative income over many years by continually adding to their portfolio. They also have to be leaders in the technology. For example, Sprague Electric's basic position in solid tantalum capacitors generated millions of dollars in income over many years. In one case where an ex-employee had stolen the know-how and set up a new company based on it, we were able to put him out of business permanently.

Patent licensing doesn't necessarily mean technology licensing. In the case of Sprague, while we would generally license our patents, we seldom licensed the related know-how itself. As noted in the Sanken story in Chapter 1, this was one of the reasons I refused to provide them our wafer fabrication expertise. In addition to a concern over creating a strong competitor, it was just plain something we didn't do. (But we made exceptions. As we'll see in this chapter, we violated this basic philosophy in a technology exchange agreement with Mitsubishi.) To illustrate the differences between patent licensing and technology licensing, let's examine some history.

After the invention of the transistor in 1948, Bell Labs' parent, Western Electric, was active around the world in licensing their semiconductor patents. Because of their basic position and the fact that they held essentially all the original patents, one could not operate without being a licensee. For example, although Sprague purchased its original transistor technology from Philco and, from a processing and manufacturing standpoint, this technology was completely different from Bell's, we still had to pay Western Electric fees on every device sold.

Licenses from Bell were instrumental in starting the Japanese semiconductor industry in the early 1950s. Although these covered both patents and technology, it is unclear exactly how much the Japanese licensees were able to obtain in detailed know-how compared to what was available in the literature and the patents themselves. While one could argue that the electronics world might be different today if Western Electric had refused to license the Japanese, they were licensing essentially anyone who approached them. (Why should they have worried about a distant, isolated country that was just beginning to recover from the war and had no position in solid-state electronics at all?)

These agreements were very favorable to Bell. As a long time licensee, my experience was that you could get nothing really useful in the way of process and device know-how unless you were as proficient in the field as they were. We used to make several trips each year to the Labs to find out "what's new in semiconductor technology." I remember asking exactly that question in a crowded conference room on the first such visit I made in the early 1960s. Fixing me with a cold stare, the Bell spokesman replied, "What would you like to know?" More memorable were the long lunches which generally lasted two to three hours. Bell Labs scientists seemed to have an enormous capacity for double martinis. As far as detailed insight into capacitor or semiconductor technology was concerned, other than through friendships that gradually developed with Bell scientists over the years, to the best of my knowledge our license agreements with Western Electric provided precious little know-how. However, they did allow us to be active in the field. As our own position strengthened, we were able to negotiate increasingly attractive cross-license agreements at lower and lower fees which never got much above 1% on sales.

There was another important point about these licenses. The Western agreements gave them access to any licensee's technology that was related to the covered field(s). In the case of Sprague this was broad coverage, including both capacitor technology and semiconductors. This access not only included our labs but also our production facilities. The Bell Labs and Western Electric scientists actively pursued this part of the terms through regular visits to Sprague facilities. We quickly became as jealous about providing them detailed information as they us. One of their most memorable visits was to Sprague's largest tantalum capacitor plant at a time when Western Electric was having severe production problems in their own solid tantalum capacitor manufacturing. At the same time they were seeking our help in solving their own problems, with typical arrogance they told us what was wrong with our manufacturing compared to theirs. What a system!

But Bell Labs was not the only source from which western and Japanese corporations gained their early start in the semiconductor field. For example, Sprague Electric entered the field in the middle 1950s by obtaining a know-how license from Philco in what was known as electrochemical transistor technology. While there were a number of different device and process variations, the basic approach was radically different from that pursued by Bell Labs and most of the rest of the industry. The Bell approach that developed after the original invention became known as the MESA transistor, because of the shape of each individual device. The key difference from Philco was that it was a batch process in which a number of transistors were fabricated simultaneously on a germanium or silicon wafer. The principle performance problem was a limited high frequency performance because of the high collector resistance. In the case of the Philco approach, each semiconductor die was handled on an individual basis and the base thickness was accurately determined by optical sensing and electrochemical etching. Then the emitter and collector contacts were directly plated on this base. Each individual semiconductor die was handled separately in some of the most highly mechanized equipment ever seen in the industry. Because of a low collector resistance, the Philco devices were the fastest switching transistors in the world at that time. This is why Sprague went the Philco route.

This was a true know-how agreement. For payments in the multimillion dollar range, Sprague purchased equipment, detailed process specifications, and engineering help. The only problem was that Philco turned out to be the wrong horse. The development of epitaxial growth at Bell in the late 1950s solved the resistance and performance problems in the MESA structure and this approach would prevail, especially when it evolved to Si planar technology announced by Fairchild in 1959.

But Bell was not the only early licenser in Japan. Several far-sighted U.S. companies used their early lead in the semiconductor or other fields to gain major positions in the infant Japanese market at a time when foreign corporations were prohibited by Japanese law from 100% ownership of local subsidiaries. For example, Texas Instruments refused to provide IC licenses to Japanese companies until they obtained permission from the Japanese government for 100% ownership of any operations they created in Japan. This was also true of IBM in the computer field. Others, such as Fairchild, looked for quick financial return and provided patent licenses and, in some cases, technology without using their patents and technology as levers to gain access to the local market. The truth is few U.S. companies in the 1950s, 1960s, and early 1970s really worried much about the ability of Japanese companies to ever become competitive with the U.S. in high technology industries such as semiconductors. They were starting too late, had too little initial capability, and, well, they were only Japanese!

It is extraordinary how much critical technology Japan has been able to purchase from the rest of the world since World War II. In *Kaisha*, Abegglen and Stalk state that between 1951 and March of 1984 Japanese corporations and institutions negotiated some 42,000 contracts for the importation of foreign technology from around the world. The total cost of this was under $20,000,000,000, a fraction of America's current annual research and development expenditures of approximately $140,000,000,000. Even today, as Japanese companies are increasing their own internal research expenditures, they still seem able to access what they need from the outside world, especially the United States. Citing just one of a multitude of examples, Dr. Denda of Konica recently told me that most of the new and innovative technology the camera company was using in a major project on electronic imaging came from several small, cash-poor American start-ups.

Even though the Japanese are now clearly a competitive threat in many of our technology industries, know-how licensing by U.S. companies continues unabated. In some cases the reason is survival—in other words, there is nowhere else to go for funds that are required to grow or to stay in business. In other cases, the main aim is quick profits. On the Japanese side, the technology is available and generally relatively cheap (especially with the devaluation of the dollar that has occurred over the last several years). Japanese corporations also are investing in our universities through such methods as membership in industrial liaison programs and consortia, direct support of professors and graduate students, and sponsoring funded chairs at such major institutions as the Massachusetts Institute of Technology.

Alliances

In addition to acquisition of foreign companies or divisions and licensing of technology, there are many variations of U.S. and foreign corporation relationships that can be best lumped under the descriptor of "transnational alliances." According to an article in the January 1, 1992 issue of the New York *Times*, some 900 such alliances or foreign investment transactions in U.S. businesses have taken place since the mid-1990s. Some were implemented by U.S. corporations as a means of accessing foreign research and development capabilities. Others, such as the Siemens-IBM and Texas Instruments-Hitachi agreements on DRAMs, the IBM-Toshiba joint project on color liquid crystal displays, and the Apple Computer-Sony cooperative arrangement in consumer electronics, combine mutual technology skills as well as spread the financial risk of the major programs involved. However, as such globalization of technology increases, many feel that the long-term result will be the acceleration of the diffusion of the United States' greatest technological strength, our lead in basic science and innovation. Because of their particular skills in technology commercialization and manufacturing, this concern is especially valid where the Japanese are involved. The continuing loss of the U.S. position in major segments of the electronics industry supports this hypothesis.

Other Sources

The United States does have sources of technology which are either less available, or not available at all, to their Asian competitors. As discussed previously, some of these are actually negative determinants relative to U.S. industrial competitiveness. We have already mentioned the ease of movement of technical personnel within an industry in the United States. Whole teams are hired away to give another company a head start—at least for the moment—or start-ups spin-off to create whole new companies and industries. The history of the semiconductor industry is a perfect example. While such rapid diffusion of technology is almost unique to the U.S. and is a rich source for growth in a virgin industry, this volatility can become a negative factor as an industry matures. This is one of the important conclusions of the MIT Commission relative to the U.S. semiconductor industry.

Many observers argue that such volatility is necessary to maintain the American lead in innovation. However, very few U.S. semiconductor companies have been able to successfully make the transition from infancy to maturity. To do so requires the ability to both create a stable environment and to maintain an entrepreneurial atmosphere that fosters continuous creativity and growth through new product generation. This is a neat trick which almost no U.S. semiconductor company, other than Intel, has been able to accomplish. If the entire top management team at Intel were to simultaneously leave and form a new company, as happened when Noyce and his associates left Fairchild to form Intel, I believe the future of Intel itself would also be in serious jeopardy. For, above all else, success in high technology in the U.S. is most dependent on the competence of its management team and leadership at the top.

Defense-related research and development in the United States is basically unavailable to the Japanese. In the past, this has been a major source of technology. This was especially true in relation to semiconductors, other electronic components, and computers following World War II. Today, however, many view such support as an obstacle to improving U.S. productivity and competitiveness in such areas as industrial electronics. Despite the recent success of sophisticated American weapon systems in the Middle East, their design and manufacture has lead to little commercial spin-off in the highly competitive industrial world where the consumer decides who wins.

TABLE 5-1

Advantages in Technology Sourcing

Japan

- access to foreign research and development
- licensing of technology developed elsewhere
- acquisition of and investment in U.S. companies
- gatekeeping
- transnational alliances

United States

- total domestic research and development expenditures
- acquisitions and mergers of domestic companies
- spin-offs from existing companies and start-up of new companies
- transnational alliances?

There is a wealth of worldwide information on technology which is available to those companies that are willing to work hard enough and, where necessary, pay to get it. Some sources are more unique to the United States than elsewhere. Compared to Japanese corporations, U.S. companies seem much less prone to use foreign, especially Asian, sources. This is at least partly due to the language problem. U.S. technology firms have also been much more inclined to license their technology and make it available through other means. While this certainly generates short-term financial return, all too often the long term result has been to create one or more very strong competitors. Table 5-1 compares the relative advantages in technology sourcing.

The Role and Management of Research and Development

The 1950s and 1960s were a period of enormous growth in the U.S. electronics industry. New companies flourished and research and development were the primary driving forces. Many, like Sprague Electric, were functionally organized and led by founding engineers who remained deeply involved in the technology end of the business. While different companies used different methods to tie research and development into the total corporate entity, the Sprague approach is an interesting example of one.

In 1959, the year I joined Sprague Electric, R. C. Sprague founded the Fourth Decade Committee with the charter to set and monitor strategic plans and directions for the company over the next ten years. All key managers were members, and I felt privileged to be added to the Committee in early 1962. We met monthly and usually spent a half to full day discussing a detailed agenda that was sent out well ahead of time to all members. Assignments were made at each meeting and reviewed religiously until complete. While many subjects were discussed, one of the primary functions was to set the goals for, and to track programs within, the centralized research and development function. The Fourth Decade Committee was the group that set Sprague's course in semiconductors—often only after violent argument—and allocated financial and human resources to the many different initiatives underway throughout the company. It was a form of consensus management that worked, even if "consensus" usually meant agreeing with "R.C." In many ways it was the direct involvement of the CEO, an engineer by training, that made the committee and the company so effective during the 1960s.

As the company grew and Sprague moved toward decentralization, the Committee ceased to exist in January of 1968 so that "R.C." could devote himself "nearly full time in the acquisition area." In hindsight, it was the decline of my father's direct involvement in the business and increasing competition from around the world that started the decline in Sprague Electric's financial and business fortunes. However, it is true that Sprague had reached a size where it had to change. It had become impossible for a few individuals, no matter how competent, to make all the decisions.

This was especially true of the Semiconductor Division. The decision to truly decentralize this business unit was one of the most important reasons for its later success. However, with the clarity of hindsight, one can certainly question whether or not extending this same organizational approach to the capacitor business units really made sense. One thing is perfectly clear: disbanding the Fourth Decade Committee without replacing it with some similar approach to overall coordination and management was a mistake. Unless a company operates as a truly decentralized conglomerate, even in large organizations with a complex portfolio of different product families, "consensus management" works. The point is that, despite how much autonomy exists within individual divisions or business units, the overall health and success of the entire company is the ultimate corporate goal. If my own experience is any measure, the alternative of hands-off management by objectives (MBO) doesn't.

As years went by, the role of the central research organization at Sprague continued to decline. By the late-1970s, it was felt to be of little use by many of the officers. For example, Peter Maden (now vice president, Solid Tantalum at STI) recently told me that, as far as he was concerned, he got nothing out of corporate research and development after the mid-1970s "because they had other priorities." With Sprague's decision to exit the ceramics business in 1990, the company also discontinued all central research and development and spun-off the laboratory as an independent contract lab, MRA Laboratories. As far as corporate management was concerned, the organization had no further role to fulfill in the company's future and thereafter all technology activities would take place at the division level. At first, I thought that the gradual disillusionment with a centralized research organization might be an isolated Sprague case. However, recent experience has proven otherwise.

All too often, research and development within American corporations is either too far removed from the business of a company, or is actually viewed as internal competition. This is less due to incompetence within the technical organization than to a lack of clear goals set by corporate management and real isolation from the rest of the corporation. The following are just a few representative examples:

- During the early part of 1988, I chaired a task force charged with reducing the time to develop new products in a mid-sized semiconductor company. This was my first real consulting assignment after leaving Sprague and ended up a disappointing failure. First, I was learning a very new role as an independent consultant after many years at the senior management level of a large corporation. I was used to setting directions and making decisions. On the other hand, as a consultant I found that success depended entirely on my ability to influence others on what should be done. Even more important as far as this particular assignment was concerned, there was no one on the committee who could implement any recommendations. It had been set-up this way with the idea that working level engineers would be more open to recommending new initiatives if there weren't any bosses present. Open they were! The task force members all identified the same two key problems: poor internal communications within the corporation, and a firm belief that the CEO was only giving lip service to new product development. While a number of other specific initiatives were also identified, the suggestion that he get more directly involved fell on deaf and angry ears. As a result, I also began the painful process of learning my new trade as one who influences rather than having direct involvement in the decision-making process, including implementation. I am told that this is how American executives are now supposed to manage. Don't you believe it! While the management process today involves considerably more persuasion than when I first entered the business world, I don't know a successful company in the United States, Japan, or elsewhere that isn't run by a strong-willed and dominant CEO.

- In another consulting assignment with a U.S. components company, central research and development viewed its charter as leading the company into new businesses. Because of this, operations personnel viewed the organization as internal competition rather than support. This situation resulted from the hiring of a new vice president of research and development who saw his role as revitalization of a stagnant organization and creation of new opportunities for the company at a time when the base businesses were stagnating. I was able to get little clarification from the top when I asked the CEO what he felt the role of research and development was. Hedging, he replied "That's a good question." No wonder there was confusion within the ranks.
- An internal start-up of a large multinational was charged with transforming a core technology developed in the central research laboratories into a new business. However, the interface between the laboratories and the new business unit were strained at best, technology transfer was painfully slow, and corporate management did little to correct the situation which was largely the result of internal corporate politics. The start-up next turned to an outside contract laboratory to help accelerate commercialization. However, it was too late. Eventually the entire enterprise was closed down as part of a corporate restructuring forced by the flagging business fortunes of the parent company. If the commercialization process had been accelerated by even six months, the start-up might have succeeded. In my opinion, it failed due to a breakdown of communications between all parties concerned and lack of clear corporate direction.
- Repeatedly I have seen central research and development groups as isolated entities and not, as should be, partners in the whole of the corporate enterprise. When business conditions worsen, such

> an environment makes the central laboratory highly vulnerable. The cause of this unfortunate situation can only be laid at the door of corporate management. The head of the labs should be chosen not only for his skills as a scientist but, even more important, as a manager and as an individual who can deal effectively with the rest of the organization. In addition, the goals of the organization must be clearly set and tied to those of the corporation as a whole. Above all, communications should be open and effective between all concerned. As we'll see shortly, the Japanese have been particularly effective at accomplishing this.

One U.S. response is to decentralize the organization, thus "bringing it closer to the action." Who then will do the research? Made in *America* and other sources note the loss of some of the great industrial laboratories, such as the RCA David Sarnoff Laboratory, and major cutbacks in basic research at such places as Bell Labs, IBM, GE, Xerox, and Eastman Kodak. Even the Pentagon, which supports roughly a third of all U.S. research and development, has been shifting its emphasis increasingly toward specific weapon systems. Unfortunately, unless the commercialization process is improved, such a trend toward trying to direct research and development efforts along more practical lines will not solve our problems.

My Japanese contacts portray a very different picture. As the United States attempts to make its research more functional, Japanese research institutions are increasing their basic research. Once almost completely dependent on foreign technology, Japanese companies are increasing their internal research and development and are now considered world leaders in such important areas as semiconductor lithography, robotics, flat panel displays, optoelectronics, advanced materials such as HiTc superconductors and other ceramics, and solid state lasers. In 1989, the top four corporate recipients of U.S. patents were Japanese companies. Internal research and development budgets continue to grow, and in 1987, nondefense research and development as a percentage of GNP was close to 2.8% in Japan versus only 1.75% in the U.S. As a specific example, at Murata fully 8% of the total workforce is involved in research and development and primarily concerned with development of downstream products.

These statistics present a grim picture and indicate that we are now starting to lose our lead where we have had our greatest strength, in basic science and innovation. Are things really that bleak? While these trends certainly cannot be ignored, research and development percentages and numbers of patents tell little about how unique and good research initiatives are. In the surprisingly critical 1991 evaluation of

Scientific Research & Education in Japan mentioned in Chapter 2, Columbia University Professor Koji Nakanishi and five Japanese academic associates conclude that, while much progress has been made, to a large degree Japanese science is still trailing the west, and especially the United States. Specific problems mentioned in Japan include the following:

- the emphasis on rote memory rather than creative thought forced by the importance of entrance exams in gaining admission to prestigious universities
- the limited number of research proposals funded
- the presence of few, if any, officials in funding agencies such as the Ministry of Education, Science & Culture who have doctorates in science or medicine
- a lack of respect for scientific contributions by fellow countrymen
- resistance to change by senior faculty at universities
- an aversion to the peer review process
- few postdoctoral fellowships
- perhaps most importantly as far as creativity is concerned, the lack of heterogeneity in research groups

Of late, some Japanese intellectuals have fueled the confrontational fires by extolling the superiority of ethnic purity while criticizing the U.S.'s "mongrel society." More realistically, Nakanishi and his associates identify the very heterogeneity and continual self-renewal of the U.S. scientific community as one of the major contributors to superior U.S. science. The point is, original science and innovation continue as major strengths in the U.S., strengths we must maintain both through continued emphasis in our academic and industrial research and by a more enlightened view of how easily we make the results available to others, especially foreign corporations. For, as discussed in the previous section, as long as it is available, the Japanese will continue to tap our core technologies. Control of this drain lies primarily in the hands of U.S. corporations which, in the past, have been the primary source of the technology outflow. On the other hand, I believe the U.S. government also has an important role to play in protection of intellectual property which, to a large degree, it is not currently fulfilling effectively.

While organizational approaches differ by company, there appear to be certain commonalties in Japan that differ from much of current U.S. practice. First, since Japan has a very limited military capability, essentially all research and development is oriented toward industrial applications. Secondly, in larger firms there are at least two levels of research and development. The first is at the corporate level and is generally oriented toward new materials, products, and processes. The second level is tied directly to a particular business unit, both in programs and in geography. Both levels pursue more practical than esoteric programs, although this is beginning to change some. What is different is the continuing movement of personnel between the different organizations and, in fact, throughout different parts of the company. The Japanese research scientist at the Ph.D. level can start just as removed from reality as his U.S. counterpart. He (this pronoun is not sexist; almost no Japanese researchers are women) just isn't allowed to stay that way. In addition, although I can quote no absolute statistics, the ratio of graduate B.S. and M.S. engineers to science and engineering Ph.D.s is much higher in Japanese industry than in the U.S. This movement of people not only creates a well-trained technical work force, it also creates linkages between individuals that help accelerate the commercialization process. Finally, CEOs of Japanese technology firms are still much more likely to be from an engineering discipline than their U.S. counterparts; in the United States, the trend is toward finance and legal backgrounds for technology CEOs. As such, Japanese CEOs are more apt to stay involved in the technology end of their businesses. As we shall see shortly, in at least one large Japanese company the leaders of technology project teams report directly to the CEO. You can imagine what kind of priority that provides each program.

In my opinion, there is an important role for a central research and development organization, assuming the CEO is committed to it, directly involved, and the firm can afford it. This role encompasses the following functions:

- a limited amount of applied research related to the corporation's business interests
- interface with the technical community and coordination of supported university research
- gatekeeping
- the development of new products, processes, and businesses
- expert consulting services throughout the corporation
- troubleshooting
- training of technical personnel (For example, the ceramics division of Mitsubishi Materials instituted a policy that every new engineering hire must spend a minimum of two years in research and development.)

There are important additional major problems that can result from complete decentralization of the technology function. Some functions, such as analytical and library services, can only be done effectively and efficiently on a centralized basis. Decentralized research and development functions tend to lack critical mass and primary emphasis tends to be on troubleshooting. Hiring top scientific talent is much more difficult. It also becomes impossible to respond to true technology discontinuities. In the case of small companies with limited financial resources, there needs to be at least one senior scientist or engineer who provides a strong gatekeeping function. In addition, some form of association with one or more technical colleges or universities as well as support of contract research, if available, should be considered.

Table 5-2 summarizes current trends in Japan and in the U.S. relative to the role and management of research and development.

TABLE 5-2

Trends in the role and management of research and development

Japan

- CEO: technically trained and involved in research and development
- tapping of foreign-sourced technology
- increasing basic research while staying close to company goals
- little military research and development
- heavy engineering content
- intracompany mobility of technical personnel
- research and development as a training ground

United States

- CEO: backgrounds tend more to finance and legal and less involved in research and development
- leadership in science and innovation
- decreasing basic research
- isolation or decentralization of research and development function
- military research and development one-third of national total
- "more practical" research and development
- little mobility
- little use for training

The Commercialization of Technology

I would like to approach this critical subject from three different perspectives: allocation of resources, teamwork in technology commercialization, and by two specific case histories that tie everything together. The second, in particular, is one of the best examples I know of to show the differences between the way U.S. and Japanese companies approach technology commercialization.

Allocation of Resources

The majority of the literature sources I cited in the previous chapter, as well as most of my direct contacts, state that process or manufacturing technology is where Japan enjoys its greatest competitive advantage over the United States. Synonymous descriptors include process and manufacturing engineering. A few specific references emphasize the point:

- *Competitive Edge* claims that Japan's primary technology contributions in semiconductors have been in process innovation, productivity, and quality.
- *Made in America* calls a lack of attention to manufacturing technology a major U.S. weakness; *Competitive Advantage of Nations* makes the same point.
- My industrial contacts at EIAJ have identified the strong position of their parent corporations and industrial groups in manufacturing technology as one of the major strengths of Japanese semiconductor companies.
- Gordon Love, Technical Director-Ceramics at Alcoa and an old friend, offers the opinion that all too often U.S. manufacturing firms have inadequate resident engineering skills and insufficient in-house equipment design and maintenance capability at their factory locations. This relates, at least in part, to his experience at Sprague in ceramics. I recently observed a similar situation at a major U.S. firm with a large central research and development organization. Because of profit center accounting, a manufacturing plant "couldn't afford" adequate in-house engineering to support a complex ceramic process originally developed by the corporate technology group.

But just what do we mean by process or manufacturing technology? A good way to answer this question is to refer back to Chapter 3 and recall the "dizzying array of things that must be understood and controlled (in electronic ceramics)." Process technology is just that—the understanding and control of the multitude of variables involved in the successful manufacture of, in this case, MLC capacitors.

Expertise generally relates to allocation of resources and all evidence demonstrates the major Japanese emphasis on engineering. We have already discussed the ratio of B.S. and M.S. degrees to Ph.D. degrees. A similar trend relates to the ratio of university degrees in the natural sciences to B.S. degrees in engineering. In 1986, in the U.S. natural sciences were favored by close to 2 to 1. In Japan just the reverse is true, and engineering degrees are nearly 5 to 1 over natural sciences. In total allocation of technical resources, two-thirds in the U.S. go into new product development and only one-third into improvement in process technology. In Japan the ratios are exactly the reverse. In-house training also supports the engineering discipline in Japan with graduate mechanical engineers often spending several years as apprentices in positions which, in the U.S., we would define as technicians. Dr. Denda's description of process control in an assembly business says it as well as anything: "I think (production engineering) must have the largest contribution to cost structure. The technology is not necessarily automation and doesn't necessarily depend on investment in expensive equipment. Engineers simply try to eliminate handling of the workpiece with human hands or fingers. They make a lot of small tools and gadgets to handle many (electronic) components at one time. They also concentrate on transporting them easily from one machine to another. This is done by skillful engineers and not scientists. They are mainly university graduates or have obtained equivalent higher education." Much the same thing is true of mechanization at Nichicon-Sprague.

Teamwork in Technology Commercialization

In addition to a heavy emphasis on engineering, the real key to the Japanese approach to technology commercialization is teamwork. Whether setting corporate goals, or operating as a project team, the Japanese spend an inordinate period of up-front time deciding exactly what should be done. Once consensus is reached, however, implementation proceeds very rapidly with everyone working to accomplish the agreed-upon goal. This is a tough process for an American. As former CEO Dick Morrison of Allegro Microsystems puts it: "In the U.S. we spend a relatively short time reaching consensus. Once reached, all too often everyone goes off in his own direction."

Technology implementation by means of cross-disciplinary project teams is prevalent in Japan. The following are a few typical examples that were discussed during my June, 1991 trip.

- At Sharp, at any one time there may be two or three major project teams. A typical project might be the development of a color LCD flat panel display. Such projects are chosen by the board of directors and the team leader reports directly to the CEO. The leader, or "champion," can choose his team members from throughout the organization—refusal to participate is not permitted—and membership is across the different company functions. Teams of 10 to 15 individuals stay together until the task is finished, which is sometimes several years, and progress is reported regularly both to the CEO and to the board. With such a membership and visibility, progress is rapid. Once the development is complete, much of the team goes to the production operation and often the team leader ends up as head of the new business activity.
- The approach is similar at Matsushita. Several such teams are generally underway at the same time and, in general, include membership from marketing, research and development, design, manufacturing engineering, and equipment design. This last point is of particular importance since this means that equipment requirements are interrelated to the materials and processes to be used during, not after, the development. Some team members may go to production once the project is complete.
- Cross-disciplinary teams are also used at Murata Manufacturing, as in the case of the development of a new ceramic sensor. Again, equipment requirements are tied in early in the development. Murata also is trying to shorten the development cycle through its "6:3:3: System," which targets no more than six months for product development, three months for manufacturing implementation, and three months for creation of the sales activity.

There are three important criteria necessary for the success of such teams. First, the champion must be a strong and effective leader and have high visibility at, and support by, senior management. Secondly, consensus on the goals must be reached early in the process. Finally, as much as possible of the development and implementation must be done in parallel and not in series.

Project teams are not a unique Japanese invention nor possible only in such a culture. The MIT Commission report mentions the success of concurrent engineering teams at IBM, Hewlett-Packard (H-P), and Ford. The July, 1991 issue of *IEEE Spectrum* was devoted primarily to concurrent engineering and included successful case histories at H-P, Cisco Systems, Inc., Raytheon, and ITEK Optical Systems. Allegro Microsystems has embraced the concept of teamwork throughout the enterprise because, states Morrison, "the IC business is just too complicated to succeed without it."

I have been personally involved in successful project teams, and also a few painful failures. The Si-planar research and development team I headed at Sprague in the early 1960s was, in effect, an example of concurrent engineering. When I went to Worcester to manage the Sprague Semiconductor Group, for the first several years the entire management group acted as a team with common, agreed-upon goals. Meeting sometimes as often as once per week, we managed the entire business through common action and agreement until financial viability had been reached. Successful, high technology start-ups are perhaps the best example of a cohesive team endeavor, as was evident in the early days of Mostek. Unfortunately, the vitality of such start-ups often fades as they grow and become successful, or are absorbed by larger corporate entities who try, almost always unsuccessfully, to impose their own culture.

Despite these positive comments, I have seldom seen the project team concept practiced in the U.S. with as broad cross-functional representation or with the discipline and attention to detail I have seen in Japan. An American executive complained to me recently that he found the Japanese almost impossible to work with because "they move much too slowly." However—more like the tortoise than the hare—they do get there. When they do, their detailed understanding of materials and processes, coupled with their continual drive to upgrade and raise standards of performance and quality while reducing cost, usually creates business leadership. The DRAM IC business is an excellent example. The two narratives that follow, admittedly widely separated in time, also vividly demonstrate the two ends of the spectrum.

Personal Case Histories

Brief entries in the Sprague Electric 1960 and 1961 annual reports tell the short history of my first new product development and introduction, the OXSIL Silicon Dioxide Capacitor. The 1960 entry read as follows:

> "The Sprague Research and Engineering Laboratories have many programs underway to investigate the basic scientific phenomena underlying our products, such as the entire field of electrochemistry and dielectric properties. One such program, in the area of silicon-silicon dioxide systems, has already resulted in the development of an entirely new type of silicon capacitor, which has unusual stability for high-temperature

> operation and which is expected to suffer minimum degradation from exposure to nuclear radiation."

The 1961 report describes the progress made during the following year as follows:

> "Our OXSIL Silicon Oxide Capacitor, referred to in last year's report, was made available in sample quantities during the year. Pilot production facilities are currently being placed in operation, and development work is continuing on improved varieties of this high stability unit, which is a product of our semiconductor research work."

A small team of us, including Warren Berner, Otto Weid, and Joe Minihan, developed the product based on an earlier announcement of a similar device by Texas Instruments. What could be better than development of a new capacitor using semiconductor technology by the world leader in capacitors? Simple in concept, it involved thermal oxidation of a roughened, high purity single crystal wafer of silicon to create the highly stable silicon dioxide dielectric layer and then evaporation of a metal counterelectrode. The oxide is highly stable over a wide range of temperature, frequency, and applied voltage, and we expected to be able to create a competitive device to the Corning Glass Capacitor which gained its stability from the glass dielectric used in their construction.

No cross-disciplinary team was involved in our program, and the entire project right through pilot manufacturing was carried out within the central research and engineering laboratories. There was little critical input from the corporate marketing group, in no small part because semiconductors were still considered an aberration by much of senior management. What coordination there was took place at the monthly Fourth Decade Committee meetings. In addition, little time was spent on a detailed cost analysis since we assumed that little problem would be the responsibility of production. And, after all, wasn't semiconductor technology supposed to be inherently inexpensive? So we concentrated on proving feasibility and on a detailed comparison of the device characteristics and reliability versus Corning. The comparison showed we had an excellent, competitive device.

Typical of the semiconductor world, the most difficult manufacturing problem related to the cleaning process prior to oxidation. We spent a great deal of time on proper surface preparation. Introduction of rigid process specifications and operator training led to a high yield, reproducible process and soon we were in pilot production. There were only two problems. First, the device had no operating characteristics that were really superior to the Corning unit. Therefore, customers saw no reason to shift unless we could demonstrate a real cost advantage. We couldn't, at least with the engineering skills we had available. No matter how we approached the problem, there seemed no way to better the pricing offered by Corning, which was based on a highly automated process. So the product quietly died and I learned an important early lesson: no product will ever replace an entrenched device offered by a competent and respected competitor unless it offers some unique advantage to the customer, be it performance, reliability, or cost. Perhaps we could have realized this in the beginning. But the 1960s and 1970s were the decades of invention, and the game was to prove something could be done and worry about commercialization later. This is still true in much of U.S. industry today.

However, the project was anything but a complete failure. During a series of experiments on the effect of the conductivity type and resistivity of the substrate silicon, we observed some unexpected results: the shapes of the capacitance and loss angle curves varied as a function of applied bias, frequency, and the conductivity type and resistivity of the substrate. Only at low resistivity levels could we obtain the flat response inherent in the silicon dioxide dielectric and which we needed in our capacitor. These observations led further experimentation and generated a number of technical papers in the early 1960s on utilization of such measurements to investigate silicon surface states and the theoretical interpretation of the results. Many were coauthored with Kurt Lehovec, and were important early contributions to the field of MOS devices. This is a direct demonstration of how manufacturing, admittedly pilot production in our case, can serve as an important source of innovation, assuming those involved have the intellectual curiosity to follow the unexpected clues.

More than twenty years later, I observed first hand the thoroughness with which Japanese companies approach the problem of technology commercialization. Unlike my previous narrative, which related to new product introduction, the program dealt with manufacturing processes and equipment. In the mid-1980s, Sprague Electric and Mitsubishi Mining and Cement Co., Ltd. (MMCC; now the Ceramics Division of Mitsubishi Materials Corp.) executed a series of agreements related to the sale of each other's products and to technology exchange in electronic ceramics, especially MLCs. No patent or know-how licensing was involved, and it was really a merging of technical resources. Sprague was looking for a way to expand the sale of ceramic capacitors in Japan and a possible future joint venture partner to help defray the spiraling costs of the business. On their side, MMCC wanted to penetrate the U.S. market and hoped to access Sprague's well known expertise in ceramic formulations. They also were intrigued by what they had heard of the Sprague "Flip Process," which might serve as an alternative to their own manufacturing approach. In addition, the technical personnel on both sides liked and respected each other, as did the management teams.

In a recent conversation with Mr. Ono, who headed the MMCC technical team, he explained why Sprague was chosen over other U.S. competitors. First, he did not want to work with someone who was much larger in ceramics, as was the case, for example, with AVX. Secondly, with the industry drive to ever smaller component sizes and greater volumetric efficiency, he felt that only a "wet" process could make the required thin dielectric layers. If you remember our discussion in Chapter 2, the thinner the dielectric layer, the higher the capacitance. The Sprague process, which he knew little about at the time, offered an interesting alternative to his own. The MMCC approach, although also "wet" and able to produce the required performance, employed a screening approach for depositing the ceramic layers. It was therefore very capital intensive. As he recalls, when he finally saw the Sprague equipment, its high throughput and low capital cost "came as a complete surprise." Less clear were Sprague's relatively low chip yields—often less than 70%—compared to industry norms of over 90%.

Suspecting that the problem lay not in the equipment but in inadequate process characterization and equipment maintenance, Ono asked if Sprague would let MMCC borrow a production Flip equipment installation so that they could run detailed experiments on it. They agreed to openly share all the information they generated with us. Jealous of what we considered to be our unique technology, and philosophically opposed to supplying know-how, we hesitated at first. However, after lengthy internal debate we finally agreed. After all, we had been unable to make any progress on yield improvement by ourselves, we might learn something useful in the process, and we had excess unused equipment capacity available. So our surplus Flip equipment was shipped to MMCC in Chichibu, Japan.

MMCC installed the equipment exactly as it had been configured at the Sprague Electric manufacturing location in Wichita Falls, Texas, and assigned three of their best engineers plus two technicians full-time to characterize the process. Most of the time was spent developing the right paint rheology for the waterfall that was used to deposit the ceramic layers and on studies of drying the deposited paints. Both were very different from their own experience. At the end of approximately one year, satisfied that they had learned all they could, they returned the equipment along with a complete English translation of all the data and conclusions. In reading over this report, I was astounded at the amount of detail and how much MMCC had learned in one year about a process that Sprague had been trying to control for close to 20 years. Unfortunately, the work was of little use to us since there were inadequate technical resources at the manufacturing location to implement the findings.

On the other hand, based on what they had learned, MMCC proceeded to develop and put into production the next generation Flip equipment. While detailed information is proprietary, I understand they can manufacture the most demanding of the MLC ratings on it with yields at least as high as industry standards, something Sprague never came close to accomplishing. In addition, in the typical disciplined and thorough Japanese approach to technology, Mitsubishi continues not only to improve the process but also to evaluate alternative manufacturing approaches.

It is evident that the net result of the Sprague-MMCC relationship was a give-away of Sprague technology, not because MMCC didn't do what they committed to but because Sprague was not organized or staffed to capitalize on the information MMCC developed and fed back to us. Nor could MMCC have developed the Sprague process on their own. This does not mean they would have failed in MLCs. Rather, they would have been stuck with a process that was much slower and more capital intensive than Sprague's. Unfortunately, our joint sales efforts were also largely ineffective. As it turned out, the product configurations and ratings differed in the served markets and neither MMCC in Japan or Sprague in the U.S. and in Europe had any real success selling the products of the other. As I now look back at the final result of our joint efforts, I should have insisted on creation of a joint venture in Japan and possibly the U.S. before, in effect, giving away our core ceramic technology. Unfortunately, hindsight is a lot easier and surer that foresight.

Why couldn't Sprague have accomplished the same result as MMCC, considering the company's many years in MLCs and the fact that the Flip process was an original Sprague Electric development? Obviously, it could have. However, most of the technical resources were in research and development working on new materials, processes, and products. The engineers and scientists in manufacturing spent the majority of their time in firefighting or, worse, in trying to re-engineer anything that came from central research and development. The problem lay not with the technical people but with allocation of resources and with management—including me. I knew that our technical resources were incorrectly apportioned and that there were severe interface problems between research and development and the ceramics business unit. But I was too busy trying to respond to the responsibilities of being the chief operating officer (COO) and then CEO, both under new ownership, and too far removed from my first love and greatest skill, the management of technology. I had to depend on other senior managers, none of whom were able to solve these problems. While this has been a Sprague narrative, my experience tells me that it is replayed all too often throughout American industry.

The point of these two stories is that Japanese success in the electronics industry is less due to a sinister conspiracy than it is to factors within each country and the way individual companies compete in the world marketplace. Major deficiencies in the U.S. relate to lack of leadership, poor corporate strategies, short term horizons, lack of cooperation at all levels, inadequate use of human resources, and, on the macroeconomic level, the high cost of capital. Despite major strengths in basic science, total research and development, new product innovation, the graduate scientific university system, and the ease of creating entrepreneurial start-ups, the United States continues to fall behind in such important electronic industries as semiconductors and capacitors.

This is less due to the lack of availability or access to technology than to how it is obtained, managed, and used. There is nothing that MMCC learned about the Flip process that Sprague couldn't have learned earlier if it had properly allocated and managed its resources. It is also interesting to speculate—idly I must admit—whether or not the OXSIL capacitor might have been a commercial success if we had been able to commit the same type of engineering resources as MMCC did to the Flip project. While not de-emphasizing the creation of basic science and new products, U.S. corporations need to provide a better balance in their overall resources and to put more emphasis on process and manufacturing technology. In addition, they need to accelerate the product development and commercialization cycle by use of cooperative efforts such as cross-disciplinary project teams.

With these thoughts in mind, let's now turn to the manufacturing floor and discuss how such initiatives should lead to a more competitive United States. This will be the subject of the next three chapters.

CHAPTER 6

The Work Force

In early November of 1989, the American Management Association (AMA) and Japan Management Association (JMA) co-sponsored a conference in Chicago entitled "Two Perspectives, World Manufacturing Outlook: American and Japanese Viewpoints." It was an interesting conference with a broad representation from different industries and disciplines in both the United States and Japan. Unfortunately, the Japanese presentations were, on the whole, hard to follow because most were given in Japanese and translated for the American audience through earphones similar to the U.N. proceedings. This surprised me since, regardless of the language difficulty, I had never heard a technical paper given in the United States by a foreigner in any language but English prior to this conference. It was just one indication at the conference of how the Japanese today view the relationship between themselves and their foreign competitors.

Several intriguing Japanese perspectives became clear during the presentations. Katsuji Minigawa, general manager of the Semiconductor FA Engineering Division of NEC Corporation, described the NEC experience with the early start-up of an IC manufacturing facility in Scotland. At one point he proudly showed a slide which gave the following as one of the solutions to some personnel problems they had in the beginning: "To Combine the Best of U.K. and Japanese Management Ways and Thus Create Harmony." During the question and answer period, I asked what U.K. management techniques he had found superior to NEC's and how he had incorporated them? After a long silence, the grim faced speaker replied, in English, "There were none!" He then went on to explain that there was only one way to do things, the Japanese (in this case NEC) way. The stunned audience sat there silently, trying to absorb the arrogance of this statement.

Shin'Ichiro Nagashima, president of Canon Virginia, Inc., gave a very similar view while describing the work force at their plant in Newport News, Virginia. Because of a very large labor pool, an exhaustive hiring process, and extensive in-house training once hired, Nagashima stated that the Virginia labor force was nearly equal to that in Japan. He said further, however, that there were much greater problems with "middle management" who just plain had trouble doing things the Canon way. The strongest criticism was leveled at what he said was management's "fairly cavalier attitude toward quality." This was also a complaint with local suppliers. Mr. Nagashima excused these deficiencies by stating they were due to poor preparation and training by U.S. employers.

I must admit that I did not find the attitude of these two speakers either terribly surprising or new, although the particularly insolent manner of Minigawa-san certainly came as a shock, especially in such a public forum. Most Japanese managers I know—both in this country and in Japan—feel ill at ease in an entrepreneurial atmosphere or in the environment of a high technology start-up. But when it comes to running an established business and manufacturing superior high volume products, to the man (again, this isn't sexist; it reflects the reality of Japanese management) they feel there is no one in the world that can do it better. In many industries—ranging from electronics to automobiles to steel—this seems to be true. But—as the United States has learned painfully in recent years—the higher you are, the further there is to fall. This Japanese sense of superiority may, in time, prove to be the same sort of Achilles heel we have had in the United States and perhaps with the same results. Waiting for this to happen, however, could take years. That is time we do not have. In the meantime, Japanese corporations do have a lot going for them that warrants their sense of superiority relative to manufacturing.

Back in Chapter 4, we looked at Japanese sources of competitive advantage which have resulted in their manufacturing superiority. One of the key Japanese advantages was identified as a highly trained work force. But how have the Japanese managed to create such a work force? In this chapter, we'll examine some reasons for their success.

Education

If education has any value, then greater relative intellectual competence and skills in a nation's work force should lead to competitive advantage. Unfortunately, regardless of the source, there is uniform agreement that the United States—despite the quality of its technical universities—has a severe problem with science and mathematics education at the primary and secondary level. Most of the books cited back in Chapter 4 support this unhappy conclusion. For example, *The Competitive Advantage of Nations* deplores the erosion of primary and secondary education in the United States and recommends local, state, and federal initiatives as well as expanded in-house training within corporations. There is strong evidence that the preschool environment is the major culprit for many children, especially the disadvantaged. Citing a survey of 7000 teachers by the Carnegie Foundation for the Advancement of Teaching, a sobering article in the December 8, 1991 Boston *Sunday Globe* by Phyllis Coons estimates that some 2,000,000 kindergarten children have severe learning disabilities due to emotional and, especially, language difficulties. While exact correlation is impossible, one must assume that these early deficiencies make the job of learning that much more difficult by the time these students finally reach high school (assuming they get there at all). National measures in math and science support this conclusion.

As specified in Sec, 4(j)(1) of the National Science Foundation (NSF) Act of 1950, the National Science Board (NSB) publishes a biennial series of reports which, among other things, show how the United States compares versus other countries in a series of national assessments of undergraduate competence in math and science. "Science & Engineering Indicators–1989" reports the results of two major such assessments, one by the International Assessment of Educational Progress (IAEP) and one by the International Association for the Evaluation of Educational Achievement (IEA). While one may question how truly representative and fairly administered these tests are, the reputations of the agencies involved validate the results and one can only conclude from the data summarized in Table 6-1 that we do, in fact, have a severe problem in our primary education system.

TABLE 6-1

Results of science and math achievement tests

Group Measured	*Measure*	*U.S. Rank*	*Source*
14-year olds (1988)	science achievement	14 out of 17 developed nations	IEA
13-year olds (1988)	math achievement	last of 12 nations	IAEP
U.S. high school seniors (1986)	biology	last of 12 nations	IEA
	chemistry	10 out of 12 nations	IEA
	physics	8 out of 12 nations	IEA

According to the IEA study, Japanese students ranked second in science achievement (Hungary was first) and well ahead of the United States in all other categories. In a special report titled "Asiapower" in the June, 1991 issue of the IEEE *Spectrum*, it is claimed that Japanese high school seniors have as great an understanding of math and science as the average U.S. college sophomore. In "Japan's High Schools," Thomas Rohlen goes even further, stating his belief that the average high school graduate in Japan has as great "basic knowledge" (whatever that is) as the average U.S. college graduate. Wow! The Japanese emphasis on statistics relative to the United States is especially important as the workplace becomes more complex and statistics is a commonplace tool in quality management and machine operations.

The causes of the problems are many and are most difficult to deal with in large metropolitan areas with their increasing numbers of minorities, single parent homes, and continuing influx of immigrants. According to Harold W. McGraw, Jr., chairman emeritus of McGraw-Hill, Inc. and currently president of The Business Council for Effective Literacy, today 60,000,000 U.S. adults can only read between the fourth and eighth grade level and half of these are in the work force. Earlier we cited the ethnic diversity of our population as one of the root causes of our entrepreneurial spirit and skill and as the source of continuing renewal within our scientific community. Yet, as indicated in the survey cited above, this very diversity also makes the education problem more difficult in the early years, especially as immigrant children with little or no capability in the English language flood urban schools incapable of handling the influx. In "Metamorphosis of the American Worker" in the November, 1990 issue of *Business Month,* Robert B. Reich bitterly describes the fate of an increasingly illiterate, non-white, non-male, and contracting U.S. work force as it is abandoned by industry in favor of an influx of skilled immigrants and off-shore labor. Ironically, foreign owners of U.S.-based corporations may provide some of the answers, as we shall discuss shortly.

Limited as they may be, there are initiatives underway to help address these problems. One of the earliest, best funded, and most successful is Head Start, founded in 1965 with Sargent Shriver as its first director. It currently serves close to 500,000 children between 3 and 5 years of age from low income families and has an annual budget exceeding $1,500,000,000. Head Start is currently one of the few programs embraced by both the business and political communities. Recognizing that much of the problem starts at home, the program requires family involvement and, as much as anything, aims at creating self-worth in children who might otherwise never achieve this important prerequisite to effective learning. Proven benefits such as a reduction in the number of school dropouts, less need for remedial teaching, and decreased juvenile delinquency have led to the recent funding of a parallel pilot program for kindergarten through the third grade called

"Transition." The initial $20,000,000 funding will cover approximately twenty different locations.

Other initiatives tend to be much more local and more modest. Project SEED was also founded in the 1960s and teaches supplementary math in the fourth through sixth grades at schools with a largely minority population. A 1988 evaluation in Dallas concluded that the program improved mathematics achievement on both a short- and long-term basis. This was also true of students whose teachers had participated in Secme (the Southeastern Consortium for Minorities in Engineering). Of more than 4000 Secme high school graduates in 1990, 85% went on to college and nearly 50% of these are in the engineering disciplines. The Family Math program established at UC-Berkeley in 1981 expands competence in mathematics by offering workshops for primarily minority parents and their children, while New York City's NESC (National Executive Service Corps) taps second career engineers and scientists as teachers.

Other, newer initiatives include SRCCF (Semiconductor Research Corp. Competitiveness Foundation), which seeks to provide teachers with direct access to industrial technology, and "Operation SMART (Science, Math, and Relevant Technology)," a recent Girls Clubs of America (now Girls, Inc.) industry-supported enterprise which is specifically targeted at introducing young girls to math and science, thus tapping an enormous pool of potential talent. The NSF has recently funded $8,600,000 to help reform precollege science education and CORETECH (The Council on Research & Technology) has recommended a five-part program beginning at the kindergarten level.

Cynical critics of such programs say, "Too little, too late!" Yet, where sufficient time has elapsed to allow true evaluation, the results all seem positive. Still, as far as coverage is concerned, with the exception of Head Start, most other programs are generally regional in nature. They have limited overall impact compared to the countrywide emphasis on the importance of education that is found throughout Japan and other Asian countries such as South Korea. One of our main problems relates to the lack of uniformity throughout the United States on the availability of tax dollars to fund education in different states and how they are dispersed. States with weak economies generally have weak public educational systems, and even in relatively wealthy states one can find wide inequities. For example, in Massachusetts public education in cities and towns is funded by real estate taxes plus whatever is distributed from the state government. As one can imagine, this leads to huge differences in the quality of education in different communities. We need to raise our accreditation standards throughout the country, provide more equitable tax treatment within individual states, and, where necessary, compliment local educational funding with state and federal funds. Having said this, if the failure of attempts to override tax caps (such as Proposition 2 1/2 in Massachusetts) are any example, taxpayer support of such increased educational funding is highly questionable.

To be fair in any comparison, it must be recognized that the education system deteriorates rapidly in Japan beyond the high school level. With the exception of a very limited number of outstanding institutions such as Tokyo University, Kyoto University, the Tokyo Institute of Technology, and Waseda and Keio Universities, Japanese college education is only average. The system chooses to produce a pool of broadly educated high school graduates, with high skills in math and science, little in the way of specialization, and to complement this with heavy in-house training and apprenticeships. The very best students attend the elite universities just mentioned or are sent abroad, generally to the United States, for advanced education. They eventually end up with a large percentage of the top management jobs in Japan. Less capable high school graduates either go on to lesser colleges and universities or directly into industry where they receive comprehensive continuing education.

There is every indication that training, and even some of the educational task in fundamentals, will increasingly fall on the U.S. corporation. Considering the starting educational level of many workers, this will be a formidable undertaking. Still, the experience of individual companies and Japanese-owned U.S.-based firms shows it can be done.

In Japan, education and training of the work force at the large corporations continues throughout the employee's career. While different approaches are used, in general such programs cover general skills, off-the-job training, correspondence courses, group activities such as quality control circles (QCC), continuous retraining, and job rotation. Such an investment is seldom lost due to the low turn-over rate. Major corporations also help with training at sub-contractors and first-tier suppliers who may be unable to afford the expense in-house. Similar, and often expanded, training is carried-out within foreign-based subsidiaries as evidenced by the presentations at the Chicago conference we discussed earlier. The following are several additional examples:

- According to an article titled "How Japanese Industry is Rebuilding the Rust Belt" by Martin Kenney and Richard Florida in the February/March, 1991, issue of *Technology Review*, more than 20% of U.S.-manufactured automobiles are now made by Japanese-owned auto assembly plants with productivity levels and quality reported to be equal to that in Japan. These are fed by close to 300 auto-parts suppliers located in the States. Total related Japanese auto industry investment in this country is more than $25,000,000,000 and represents over 100,000 jobs. Who says the U.S. work force can't compete?

- NUUMI, the Fremont, California joint venture of Toyota and General Motors, has successfully transplanted Japanese management, manufacturing, and training techniques to become an effective worldwide competitor.
- Jack Driscoll of Murata Erie North America (MENA) says that the quality and productivity of their U.S. operations equal those of the parent company in Japan.
- Bob Yoshida, who helped set-up several of Matsushita's U.S. electronic component operations, claims that the U.S. labor force is "just as good" as that in Japan. It does, however, require very heavy in-house training "starting at the management level."

Life in these Japanese-controlled U.S. plants is not easy. As in Japan, the work ethic is demanding, there have been confrontations between labor unions and management, and some employees resent the occasional methods of intimidation used to create company loyalty. Although women are very much second rate citizens in Japan—especially as far as the work place is concerned—I have seen little carryover of this into the U.S., except their complete absence in the executive suite. (However, this same statement can also be made of a large percentage of U.S. corporations.) The American worker tends to be more independent than his Asian equivalent, and creative U.S. engineers often rebel against the rigid Japanese management style and seniority system. No matter how competent, the senior manager—almost always a Japanese transplant—is often resented for his "foreigness," at least in the beginning. Yet, the work ethic I have seen in successful electronic component manufacturing plants in the U.S. run by U.S. managers can be just as demanding. There is no evidence that the U.S. worker doesn't want to work, as long as he or she is treated fairly, paid a decent wage, knows what the job is, is capable of fulfilling it, and, above all else, feels that the company is successful and will remain so. The big issue today in the U.S. is job security, a perceived advantage as far as employment by Japanese subsidiaries is concerned, and which is worth a lot of aggravation. Unfortunately, it is currently a rare commodity within U.S. corporations.

As far as U.S. engineers and other white collar workers are concerned, those I know who work for Japanese subsidiaries in this country not only speak highly of their Japanese bosses, they are very hard to hire away from them once established. Again, job security is the major issue. For example, the U.S. employees of Allegro Microsystems (which, incidentally, is still run by an American) feel much more comfortable as part of Sanken Electric than they did in recent years as part of Sprague Technologies. In the mid-1980s, I tried to hire Jack Driscoll back from Murata-Erie, a move which he refused to make because he felt he had been more than fairly treated by his Japanese boss, Fred Chanoki, and the Muratas. He made his decision while realizing he could never become boss of the U.S. subsidiary.

The U.S. Manager

Before proceeding further, we need to discuss how management figures into the broad context of industrial competitiveness.

There are no real statistics or measures that effectively allow comparisons between U.S. managers and their Japanese counterparts. The best of each are certainly equals. However, the trends in the electronics market indicate that the U.S. definitely has problems and deficiencies and, by inference, so do the CEOs of troubled companies.

From the ivory towers of academia, a management consulting firm, or a corporate research center, it is easy to have all the answers on how a plant, division, group, or business should be run. That is exactly how I felt when, in 1968, I went from senior vice president for research and development at Sprague Electric's corporate research center to take over responsibility for the semiconductor division. It took me about two days to realize that, other than a deep understanding of the technology, I knew nothing about running the business, or, in fact, any business. I had never had to meet payroll, or respond to an enraged customer faced with a production line shutdown due to late delivery, or deal with a quality problem on a newly released IC, or explain why I (not "they") was losing money, or balance all the complex human psyches involved in a semiconductor activity. The only good thing was that I realized how little I knew, wanted desperately to succeed, and, over time, was able to surround myself with people who were smarter than I in all those missing skills.

The point of these opening comments is that it is a lot easier to know what to do when it isn't you that has to do it. This is especially true, I feel, of the U.S. senior manager or CEO who faces a myriad of pressures not necessarily felt in other countries, including the unending pressure to do better every single quarter, no matter what. Failure to do so these days may mean not just loss of one's job, but even loss of the company.

In our discussions in previous chapters, the average U.S. manager (if there is such a thing) hasn't come off very well. The portrait in a number of previously quoted texts is of an individual who cares primarily for himself or herself, takes a short-term view of the business, has difficulty in motivating employees and providing leadership, is removed from direct contact with both customers and suppliers, and neglects human resources. For example, *The Competitive Advantage of Nations* by Michael Porter points out the critical importance of the CEO to corporate success by stating one reason for the slide in the competitive strength of U.S. corporations is due to less technically trained managers and the greater emphasis on numerical results than on technology or manufacturing.

From statements such as those by the speaker of the Japanese House of Representatives, Tadahiro Mishima (concerning "inferior, illiterate U.S. workers"), and similar comments concerning poor American work habits by Japanese Prime Minister Kiichi Miyazawa, one might suspect that the Japanese hold U.S. blue collar workers in universally low esteem. Yet such comments are not supported by Japanese managers of U.S.-based subsidiaries of Japanese companies. These managers tend to praise the U.S. blue collar work force—if properly screened and trained—while being much more critical of their U.S. managerial counterparts. A particularly inflammatory statement by Shintaro Ishihara of the Liberal Democratic Party in Japan blamed the United State's industrial woes on shortsighted and self-centered management. A similar theme was introduced in *The Japan That Can Say "No,"* a book he co-authored with Akio Morita of Sony Corporation. An August, 1990 article in *Business Month* entitled "Japan's Bosses in America," states that one reason there are virtually no U.S. heads of Japanese subsidiaries in America is due to the difficulty the Japanese have in hiring and keeping the best U.S. managers. The reasons given are compensation, lack of proficiency in the Japanese language, and their feeling of exclusion from key decisions. Since most of the key shots are still called in Japan, the last is probably the most important. In addition, with the Japanese confidence in their own management practices, it is a rare American indeed that can make it to the top of a Japanese subsidiary.

Within most large Japanese corporations, there is an unwritten rule that foreign subsidiaries are to be headed by Japanese managers that come from the parent. But this is not always so. At the time this book was written, Marshall Butler and Dick Rosen were CEO and president/COO respectively of AVX Corporation, a Kyocera Group Company, and Allan Kimball was president of Sanken-owned Allegro Microsystems, having replaced Dick Morrison as head of the U.S. subsidiary. Only time will tell whether these organizational structures continue.

So just how good—or bad—is the U.S. senior manager? Like any community of individuals, it is impossible to draw a composite picture that covers the whole distribution. In the components industry, most I know are intelligent, competent, hard working, and care passionately about their companies. This is especially true when they have worked for only one or two companies and know the business inside out. This comment is less applicable in the case of many conglomerates where the top person is really an asset manager and not involved or knowledgeable in any specific business. (I must admit there is a good deal of personal prejudice in these observations.)

Companies, not nations, compete. Since the CEO or senior manager is probably the single most important factor in success or failure, the good ones are invaluable. In Japan, movement from one company to another is rare and the rise to the top invariably comes from within a firm, usually starting with a strong hiring program directly out of the university system such as the Murata "freshman program" discussed earlier. In addition, there generally is a strong seniority system which can lead to frustration on the part of a fast-track young executive who finds his way to the top blocked by a more senior but less competent manager. One would think that this could lead to stagnation, and occasionally it does. Overall, however, the system seems to work. I observed this first hand on more than one occasion in Japan where a junior man had the actual operating responsibility while the senior manager was, in effect, a figurehead with a face-saving title. As already discussed elsewhere in this book, top Japanese managers are well trained, know their business inside out, are extraordinarily competitive, and live to work and to succeed. On the flip side, they also tend to fail as entrepreneurs and—as noted several times previously—top women executives just don't exist. The system, at least up to now, doesn't allow them in the executive suite. (But anyone who doesn't think this is also true in most of corporate America has been wearing blinders.)

Many successful U.S. corporations follow the same route with top executives promoted from within after many years of learning the business from the ground up. On the other hand, seniority is less strictly followed. Often promotions come at the expense of others with longer terms of service. I believe that successful electronics firms—for example, AMP, IBM, Motorola, and Intel—have all tended to groom and promote from inside the company. Interestingly, this has also been the general rule with the U.S. auto makers, with a few well publicized exceptions such as Lee Iaccoca's move from Ford to Chrysler several years ago. In the case of autos, however, parochialism resulting from excessive inbreeding has been identified by critics as one of the major causes of today's malaise within the industry. I am not really sure why a system that has worked so well throughout Japanese industry seems to have failed in Detroit. Perhaps it is related to poor selection of leadership by the board of directors of such companies.

On the other hand, U.S. companies in trouble all too often look outside to find the new CEO. In many cases, they search for "the best athlete" regardless of whether or not he or she knows the business. While this can and has worked in many cases, it is fraught with peril no matter how successful the candidate's previous track record. Success in one environment is no guarantee the same will happen in the new position, especially if the culture is very different. In the beginning, the new executive invariably seems to lack the warts, blemishes, and faults everyone has once you get to know them. There is also a whole process of learning the business to go through, and both these factors usually provide a honeymoon period with the board and stockholders that an insider seldom gets. Finally, as an outsider he or she will often bring in their own team, further upsetting the internal balance that existed before the change. If the new manager is from a different industry, so probably are they. Therefore, unless the situation is so far gone that radical change is the only answer, such a series of changes can easily kill the patient rather than save it. Such an approach is almost never used in Japan.

There is another difference between the U.S. and many other countries, including Japan. Most U.S. technology companies were founded and led by technically trained people. However, there is an increasing trend to turn the leadership over to individuals with either a financial or legal background. While such training certainly doesn't guarantee failure, it does seem less than appropriate in today's highly competitive environment where the battle is being fought primarily in the laboratory and on the manufacturing floor. We are also a nation that loves fads, and during the early to mid-1980s graduate business schools and the MBAs they churned out were a big fad. Many of these schools are now viewed as archaic and offering precious little to help U.S. competitiveness, since most of their graduates were skilled in areas such as finance and administration rather than industrial management. The manufacturing floor is where the action is today. The problem is that manufacturing management skills must, to a large degree, be learned on the job. Starting salaries on this tract are in a different cosmos compared to the brokerage and mergers and acquisitions

("M & A") fields. Yet the pendulum does go back and forth. If job opportunities end-up swinging toward technology and manufacturing, that is where new graduates will have to turn.

Management of Japanese-Owned American Companies

What kinds of managers have the owners of Japanese subsidiaries chosen to run their operations in the U.S.? First of all, with certain rare exceptions they are invariably Japanese with many years experience in the parent company. The August, 1990 issue of *Business Month* carried an article titled "Japan's Bosses," which profiled several Japanese managers of U. S. subsidiaries of Japanese firms. Table 6-2 gives a quick summary of some of the most significant characteristics of those managers.

TABLE 6-2

Characteristics of Selected Heads of Japanese Subsidiaries in the United States

Name	*Age*	*U.S. Subsidiary of. . .*	*Background*	*To U.S.*	*Other Foreign Assignments*
Akiya Imura	40s	Matsushita Electric	English literature; "diplomat"	1987	United Kingdom
Masaaki Morita	69	Sony	Electrochemistry; "factory man"	1987	none
Yukiyasu Togo	50s	Toyota	"salesman"	1983	Thailand, Canada
Kiyohide Shirai	49	Kyocera	Ceramics, "technologist"	1971	none
Sadahei Kusumoto	64	Minolta	"salesman"	mid-1950s	none

While these managers all share the common Japanese characteristic of one company business experience, they certainly aren't clones of each other or of some imagined "Japan Inc." executive figure. Rather, like their U.S. counterparts, they are highly individualistic. Some, such as Akiya Imura of Matsushita, are quiet men sent to pour oil on turbulent waters. While maintaining very close ties to the Japanese parent (founded and headed by his outspoken brother, Akio), Masaaki Morita of Sony operates more like an American executive jetting around the country on the fast track. Toyota's "Yuki" Togo is the consummate salesman, while Kiyohide Shirai tries to extend the strong Japanese culture of Kyocera founder, Dr. Kazuo Inamori, to his U.S. operations. "Sam" Kusumoto of the camera giant Minolta has been in the U.S. so long he seems more American than Japanese. Nor are they all technically trained. For example, Akiya Imura of Matsushita majored in English literature.

There are some commonalties, however. They know their businesses. While stating they want to meld the best of American and Japanese management techniques, they invariable lean toward the latter. They like the United States, try to "act American," but are also politely critical of many of our business techniques, including the quality of U.S. goods. For example, Kusumoto-san of Minolta came to the U.S. in the mid-1950s, likes American goods for their engineering and design, but levels three criticisms at U.S. managers: poor communication with the work force, inappropriate compensation, and lack of flexibility.

Several observations stand out in these portraits. First, the parent companies and their managers have succeeded in a foreign land through years of hard work and commitment, as has been true of successful U.S. companies in Japan. Secondly, they have done it by adapting their techniques to our environment without, however, changing their operating philosophy. Third, they have succeeded while still operating under strong control from their Japanese parents, not an easy task for such ambitious men. Finally, they probably would succeed as managers of any U.S. company in the same field, if given the time to create their own management style.

It is interesting to note that, unlike the Japanese approach, heads of successful foreign subsidiaries of U.S. corporations are more often than not headed by local managers than by U.S. expatriates. This is by choice rather than by local government edict. Sprague Electric has tried it both ways. In Japan, I believe it is far preferable to have a local manager. This is also true of Europe. On the other hand, where the operations are primarily oriented toward low-labor cost manufacturing—such as electronic component assembly and test—the best executive with the right background and experience should be chosen, regardless of origin. Sprague successfully operated such factories for many years in the Philippines, Malaysia, Taiwan, and Hong Kong using American expatriates. As far as Japanese subsidiaries in the U.S. are concerned, I believe that over time the Japanese will begin to grudgingly adapt the philosophy of using experienced American executives (admittedly on a very selective basis).

Managerial Compensation

One of the greatest areas of criticism leveled at top U.S. managers relates to their compensation which, all too often, seems to have no relationship to performance. Horror stories abound of executives receiving large raises and bonuses during periods of major layoffs, restructurings, and losses. According to an article in the April 15, 1991 issue of *Industry Week* titled "CEO Pay" by Joani Nelson-Horchler, the CEO of an average mid-sized U.S. corporation receives about $2,100,000 in total annual income while, for large corporations, this number rises to $3,300,000. These amounts include base salary, bonuses, stock options, and other types of long-term compensation. Such remuneration is at least several multiples or more that of an equivalent Japanese manager. Heads of Japanese subsidiaries in the United States sometimes find themselves compensated below some of their American subordinates and have difficulty understanding why compensation schemes are so different over here. Certainly this has caused problems when purchase of an existing U.S. company or business unit is concerned, especially if a Japanese manager is transferred to run it.

Obviously direct wages, bonuses, and options tell only part of the story. Japanese executives generally have fat expense accounts, company-paid country club and other types of memberships, subsidized housing, and other attractive perks. So do U.S. executives. For example, a direct comparison of U.S. and Japanese executive compensation packages in the auto industry was included in the article "Motown's Fat Cats" by Thomas McCarrol in the January 20, 1992 issue of *Time*. In that article, it was shown that Chrysler's Lee Iacocca received compensation of $4,500,000 the previous year, and perks such as use of a corporate jet, legal counseling, a new car every year, special parking at corporate headquarters, a New York hotel suite, and stock incentives. By comparison, Shoichiro Toyoda of Toyota was paid just $690,000 and his perks consisted only of a golf club membership and shared season tickets to various events for the purpose of entertaining clients.

Further insight into the entire issue of relative compensation can be found in an excellent report released each year by Towers Perrin. Through the courtesy of Wolfgang Glage, vice president of TPF&C, I was able to obtain a copy of their "1991 Worldwide Total Remuneration" report. The data compares compensation of corporations with annual sales of approximately $250,000,000 in 21 countries and is full of good information. This report also includes in total remuneration the cash equivalent of executive perks. In the case of Japanese and U.S. CEOs, these equivalents average roughly $70,000 and $25,000, respectively. So, in general, perquisites are much more important in Japan than in the United States. I have summarized key parts of the Perrin report as they relate to Japan and the U.S. in Table 6-3, below. The U.S. compensation numbers, while high, are more in line with my own experience and considerably lower than those quoted in the Joani Nelson-Horchler article above.

TABLE 6-3

Relative compensation between the United States and Japan

Job category	*Country*	*Total Pay*	*After Tax Pay*	*Net Purchasing Power*	*Retirement Income as % of Final Cash Pay*
Factory worker	Japan	$33,000	N/A	N/A	65%
Factory worker	U.S.	$30,200	N/A	N/A	96%
Accountant	Japan	$37,900	N/A	N/A	53%
Accountant	U.S.	$54,100	N/A	N/A	77%
Human resource manager	Japan	$183,200	$90,000	$20,000	41%
Human resource manager	U.S.	$168,500	$82,000	$32,000	54%
CEO	Japan	$371,800	$137,000	$31,000	34%
CEO	U.S.	$747,500	$260,000	$100,000	51%

There are a number of interesting comparisons in Table 6-3. For example, U.S. factory workers now receive lower compensation than their Japanese equivalents. They also fall below those in Switzerland, Germany, Sweden, Canada, Belgium, and the Netherlands. However, upon retirement they get to keep a much larger percentage than the Japanese worker as retirement income. This is true in all four employee categories. There are disparities among salaried professionals. Accountants in Japan are paid less than in the United States, while the reverse is true of human resource managers. And U.S. CEOs receive compensation equal to approximately 25 times that paid to factory workers. In contrast, Japanese CEOs receive compensation approximately equal to 11 times that paid to factory workers.

By any measure, U.S. CEOs receive total compensation far in excess of their Japanese counterparts or the CEOs of any of the other 21 countries. This is the result of both higher base salaries and—especially—long term compensation. The ratio of total compensation to that of the factory worker is more than twice that in Japan (that is, 25 times versus 11 times). However, it is still a much lower multiple than in Venezuela, Mexico, Hong Kong, Brazil, and Argentina. Most startling is the relative buying power between the U.S. and Japanese CEO. Normalized to a factor of 100 in the U.S., and taking into consideration the relative cost of living related to purchase of goods and services (not, for example, housing), Japanese CEOs have less than one-third the purchasing power of their U.S. equivalents and no more than that of the U.S. human resource manager. Only in Mexico does the CEO have purchasing power approaching that in the U.S.

From personal experience, I believe that at least part of the U.S. problem stems from the hiring of managers from outside the company. For example, when trying to build a new management team at Sprague in the early 1980s, and having decided that several key positions could not be filled from within, I found that many of the potential candidates were more interested in how they would be compensated than in what they were being hired to do. As a director of another company where a retired CEO had been replaced from the outside, I discovered the board spent most of its time trying to design innovative incentive systems for the new team where before they had concentrated on how the business was being run. In still another case, when the CEO of an electronics company failed to earn his bonus due to the cost of a restructuring reserve, he received it anyway when the compensation committee argued that he had done the responsible thing, despite the negative impact on his compensation. I used to believe that CEOs were hired to manage a business for the benefit of everyone—not just themselves.

So who is at fault? In fairness, the U.S. CEO has one of the highest risk jobs in the world. This is due, in large part, to the never ending pressure for continuing improvement in short-term performance that comes from the board of directors and, through the board, the investment community and stockholders. In today's highly volatile economic and business environment it is no wonder that, under such pressure, the CEO is often forced to give "number one" first consideration. If a CEO is being hired away from a good position in another firm, then about the only way to get that person is to pay what is required, including providing stock options, long-term incentives, and the necessary perks. In order to reduce financial risk, a termination agreement or "golden parachute" is also often required. However, do this for the new person and you are invariably forced to also do so for members of the new team and any key existing executives who are survivors.

In my opinion, much of the fault lies with the board of directors and its compensation committee. With a start-up or smaller company, while the financial return may be limited, a director is still often called upon to provide real support and advice, contributing experience in a variety of ways. This is much less true with large, well-structured corporations where the associated prestige and compensation for a director can be very large. Including retainer and meeting fees, annual remuneration for a board seat can be between $50,000 and $100,000, or even higher if there is also a consulting agreement. This can be very important financially to a director. Stockholders have the responsibility for choosing the members of the board, although this is usually a rubber stamp process if there are no large concentrations of equity in a few hands. The board, in turn, picks management. There is an inherent conflict of interest in such an arrangement. If the board brings in a new CEO, that person is "their boy" or "their girl" and for some period of time will find very strong support from those behind their selection. As turnover occurs among the directors, the CEO will attempt to stack the board with individuals who, he or she believes, will give support. (During an interview with a potential CEO candidate, I asked what type of support he would hope to receive from his directors. With rare honesty he replied, " a rubber stamp." Not a very smart reply, but certainly the truth.)

For all these reasons, and because of the money involved, there is a very strong tendency for the board to support the CEO until it is obvious that things are really out of control. By then, it is often too late. Looking at the dismal results of all too many U.S. corporations today, one can only conclude that many boards are not doing their jobs properly. Having drawn this conclusion, it is also true that, because of the increasing frequency of stockholder suits, U.S. boards of directors today are being forced to become more responsive and responsible than in the past.

There are changes that would make sense. Management bonuses should be based on total company, or business unit performance, while subjective measures such as performance against personal goals should be deemphasized. Long term incentive programs should not allow any participant to begin to cash-in for at least five years or even longer. Management guru W. Edwards Deming would eliminate all bonus and pay-for-performance plans other than profit sharing. Deming argues that it is not possible to measure individual performance due to the unmanageable influence of the economy and the cooperative nature of any successful business activity. In addition, he states flatly that people are not motivated by compensation anyway. (Many U.S. executives—and ordinary employees—would flatly disagree with this conclusion!) Compensation is certainly only one part of what really creates motivation, but in times of economic uncertainty it is a very important part. In addition, trying to hire and keep executives with a non-competitive compensation package just won't work—that is, unless all corporations are willing to simultaneously level the playing field!

Training

Management effectiveness is only one part of the equation that determines the competitiveness of a firm. How well the workforce is trained is another important factor. Unfortunately, this is another area where the United States lags behind Japan.

Many reasons are given for the relative lack of effective in-house training by U.S. firms. These include a short-term perspective, financial considerations, and concern over loss of the investment due to turnover. While this last reason is the one most cited, I really believe all too few U.S. managers feel that training is very important. However, like the concept that "quality is free," a well trained employee will make—not cost—the company money. The work will be done with higher productivity, the resultant product or service will be of higher quality, and, in general, turnover will actually be lower. All these factors represent long-term cost savings. In addition, in today's increasingly sophisticated manufacturing environment, without adequate and continuing training there are an increasing number of jobs that just plain can't be done. Imagine, for example, sending an employee into the wafer fab area of an IC company to operate a $500,000 lithography system or ion-implanter without adequate training and understanding of the systems. Or the manufacture of high performance MLCs on sophisticated equipment by untrained operators. And just try to implement and effectively use SPC (statistical process control) on the production floor with inadequate background and training! The list of examples is endless.

Many training approaches are not only inexpensive, they also provide added capability during the process. Examples include on-the-job apprenticeship training programs involving students from local high schools or trade schools and "co-op" programs with nearby schools and colleges. Such programs provide both hands-on work as well as an excellent method of early evaluation and screening of potential new employees. If eventually hired full-time, these early work experiences usually create extreme company loyalty in the employee. An individual's first job should be one of the most informative experiences of a lifetime.

Many successful companies, both large and small, have identified training as a critical component of survival in an increasingly competitive world. Those that can afford it carry the process beyond training in specific, required skills all the way into higher education. Motorola has been a leader in training its workers, at an annual cost of nearly $50,000,000. Included in the extensive agenda are joint programs with local schools and community colleges. According to the October 15, 1990 issue of *Industry Week*, a number of "America's Best Plants" have extensive employee training programs that emphasize cross-training, teamwork, heavy employee involvement, and continuous improvement programs.

When Sprague Electric was a major employer in North Adams, Massachusetts, it had extensive cooperative programs with all the local schools and colleges, including a masters program with Williams College in math or science which could be earned while on the job. There were also co-op programs with Northeastern and North Adams State as well as financial support at major universities for Ph.D. programs. Several of the company's top researchers received their entire post high-school education, including the Ph.D., through this process. Peter Maden, with extensive experience in Japan, states flatly that the Sprague Electric Tantalum capacitor plant in Sanford, Maine is second to none in the world. In no small part this is due to its educated and highly trained labor force. The consistently superior performance of the plant, year after year, supports this claim.

Even some successful, smaller firms complement their in-house training programs with supplementary education initiatives. For example, Aerovox, Inc., a capacitor manufacturer located in New Bedford, Massachusetts, with annual revenues in the range of $75,000,000 offers an in-house program whereby employees can earn a high school graduate equivalency degree (GED) or certificates in math or other adult education courses.

The point is, industry must share the educational load and take the lead in training its workers. The "we can't afford it" argument just doesn't wash. As far as I am concerned—for all the reasons already stated—we can't afford not to. That doesn't mean that all companies must or can afford to offer support of higher education. However, if the funds for adequate and continuing operator training are not or cannot be made available, then the only other options in the high technology field are to either become purely a distributor of products and services offered by others, or to close the doors. If a company is located in an area where the labor pool is saturated and turnover is very high regardless of how companies treat their employees, it is time to relocate.

These comparative problems with education and training don't relate only to the Japanese. We can take little solace in the fact that the South Koreans hold education in even higher reverence than the Japanese, or that Hong Kong, followed closely by Singapore, received the top scores in chemistry and physics proficiency at the senior high school level in the IEA assessment exams summarized in Table 6-1. As we will soon discuss in more detail later in this book, the worldwide competitive race in technology is not just between Japan and the United States.

Labor Relations

Another competitive advantage enjoyed by Japanese corporations relative to those in the U.S. is a much more cooperative relationship between management and labor. This is largely due to the structure of the Japanese society which is built around loyalty to the group, which in this case is the employer or company. Since unemployment has traditionally been almost non-existent in Japan, and "lifetime employment" is practiced in the large corporations, there is often really not a lot of reason for confrontation. However, it does occur on rare occasions.

I was in Tokyo in the middle 1980s at a time when several semiconductor companies had to institute temporary layoffs due to a severe cutback in world demand for DRAMs. Workers were marching in the streets close to the office building where I was meeting, and I had a chance to see a labor demonstration in Japan for the first and, incidentally, last time. It was more like a colorful parade. Both sides of the street were lined with armed policemen, and the marchers proceeded with almost military precision down the center of the street. They seemed to be all male employees and their heads were ringed by colorful bandannas. Trucks with public address systems led each group, apparently spouting their complaints. It was better organized and controlled than most Labor Day parades in the United States, and was over about as quickly. I must note, however, that my Japanese hosts were very embarrassed by the entire affair and were quick to assure me that this occurrence was a first. I wondered about that statement.

According to Abegglen and Stalk in *Kaisha*, large Japanese corporations have what are known as "enterprise unions." Such unions cover about 30% of the labor force, compared to around 20% in the United States. The U.S. number is surprisingly low and points out that the American stigma of confrontational labor and management relations covers a relatively small percentage of the total work force. However, these tend to be in highly visible and troubled segments such as the automotive industry.

The enterprise union of a Japanese corporation includes all employees up to the management level. Because company executives invariably come up through the ranks, many Japanese CEOs were, at one time, union members. (Try to find a parallel to that in the United States!) Thus, union membership creates no barriers to job mobility within the company. This also means there is a single bargaining unit, or *shunto*, representing the employees at the annual wage negotiations held each spring. Because of extremely limited mobility between companies, the fact that each union is unique to a single company, and the practice of lifetime employment, wage settlements and other employee-management issues are negotiated so that the company's fortunes are not damaged. In other words, the needs of both groups are only served if, as a result, the corporation remains healthy. However, as the demonstration I witnessed shows, this cooperative relationship could change to confrontation if the business environment necessitated major layoffs or violation of the lifetime employment concept.

As one might imagine, when Japanese subsidiaries are located in the United States or other overseas locations, they have an entirely new learning process, especially if required to deal with large national unions such as the United Auto Workers. To the degree possible, they have mitigated this problem by trying to locate in areas where there is an excess labor pool, no strong history of union representation, and by instituting exhaustive screening and training processes. Once established, they work extremely hard to keep unions out and, where they exist, to maintain good cooperative relationships. Not surprisingly, the labor situation is an important variable when Japanese companies evaluate the possible acquisition of a foreign firm.

The history of labor relations at Sprague Electric provides an interesting example of the difference between something close to an enterprise union and representation by a national affiliate. By coincidence, my brother, Bob, spent much of his business career in industrial relations for Sprague Electric until his retirement in 1980. (Unfortunately, Bob was tragically killed in 1987 while piloting a high performance Christen Eagle II aircraft.) Sprague Electric was non-union when it was founded as the Sprague Specialties Company in 1926. In the late 1930s, its employees at the North Adams, Massachusetts facility sought and received representation by three different unions: the Independent Condenser Workers Union, Independent Office Workers Union, and the International Association of Machinists (an affiliate of the American Federation of Labor). I feel this happened because Sprague was going through a difficult financial period at the time and the employees felt there was greater strength in group representation.

The first two independent unions were by far the largest and, with the exception of a minor labor stoppage in the late-1930s, worked closely with Sprague management for many years much like the enterprise unions of Japan. Unfortunately, this also led to a perception by some employees that the unions were in the company's pocket. I believe this facilitated later take-over of these unions by national unions. Except for several overseas plants that were built after World War II and one U.S. plant that was acquired, no other Sprague locations were, or ever became unionized, although during the rapid growth period following the World War II there were several related campaigns and unsuccessful NLRB (National Labor Relations Board) elections in different U.S. branch plants.

This peaceful relationship became increasingly strained as the company fortunes continued to improve and Sprague employment in North Adams began to exceed the immediate labor pool. As a result, Sprague was forced to go further afield to fill the North Adams openings and the nearby towns of Adams and then Pittsfield became increasing sources of labor. This was also IUE (International Union of Electrical Workers) country. Many new employees came from families where other members worked for General Electric in Pittsfield, which was already well represented by the IUE. (I am not implying the IUE had been standing idly by as Sprague continued to prosper!) Sprague was a juicy target and, after several unsuccessful campaigns, an NLRB election was held in 1966 and the IUE took over representation of the Condenser Worker Union. The Office Worker Union chose to join with AFTE (American Federation of Technical Employees). As a result of these two elections, Sprague's entire North Adams blue collar force came under representation by national affiliates. After long and difficult negotiations, three-year wage and benefit contracts were finally signed in early 1967 with all three unions.

In the period of relative labor peace that followed, employment in North Adams dropped gradually from a high of more than 4,000 employees to just over 3,000 as Sprague expanded in such locations as Wichita Falls, Texas, Worcester, Mass., Hillsville, Virginia, and Lansing, North Carolina. Perhaps the national unions felt threatened in February, 1970, when negotiations over the new contracts broke-down. Led by AFTE, all three unions walked out. The result was an ugly work stoppage from which neither side ever quite recovered.

I discovered that the big issues in a labor conflict can just as easily be personal and emotional as economic ones. Since I was then located in Worcester, I only occasionally had to cross the picket lines strung around the Marshall Street complex in North Adams. Still, the pushing and shoving and strident cat-calls from people I knew left a bad taste that never quite disappeared. The local newspaper, the North Adams *Transcript*, only added fuel to the fire by openly supporting the unions and attacking Sprague.

The strike did more than cost North Adams jobs; it almost destroyed Sprague Electric. It occurred during one of the industry's periodic recessions and at a time when Sprague was beginning to experience increasing competition from Japan. Although less than 15% of Sprague's total production was affected, our competitors were successful in convincing many customers that the stoppage would spread to all of our U.S. manufacturing locations. By the time the strike was settled

10 weeks later, Sprague had lost close to 50% of its U.S. capacitor market share. It took years to recover.

Through the 1970s, we continued to move jobs out of the area to other non-union plants. North Adams had become our most expensive manufacturing location. In addition, North Adams was no longer the geographic center of Sprague's operations, which had now spread all over the United States. Add to this the fact that North Adams had never offered easy access to our major customers and there was little reason to remain there other than history. Because that history was so strong, (and because the Berkshires had been my family's home for so many years), I tried to rejuvenate the location when I returned to headquarters as President and chief operating officer in 1977. Even if North Adams could no longer serve as a primary manufacturing location, I believed, perhaps it could continue as a corporate and technology center.

Unfortunately, it soon became evident that North Adams was the wrong place at the wrong time in the company's history. Finally, under pressure from Penn Central (PCC), which had acquired Sprague in 1981 as part of G.K. Technologies, we moved corporate headquarters to Lexington, Massachusetts in 1984. This was a far superior geographic location as far as access to the branch plants and customers was concerned. The new Route 128 location also projected the high technology image Penn Central was seeking. Although not my doing, it has since moved again, this time to Stamford, Connecticut as headquarters for Sprague Technologies, Inc. Today, only a few hundred Sprague employees still remain in North Adams.

Could the painful 1971 strike have been prevented? John Winant, who worked closely with my brother in Sprague's industrial relations department, thinks not. If local employment had been kept at much lower levels—say no more than 2,000—it is possible the national unions might have failed to gain representation. Once the IUE was in, however, I feel it was only a matter of time before they had to demonstrate how much the employees had gained by joining them. On its part, Sprague could just not afford the General Electric levels of wage and benefits demanded by the IUE. If only North Adams were involved, it might have been affordable. However, we were convinced that whatever happened in North Adams would eventually be required in all the branch plants, and that we couldn't afford. As a result, no one won—least of all the workers who eventually lost their jobs. For, if actions are taken that destroy a company's competitiveness—whether by labor, management, or both—ultimately there are no jobs for anyone. As the former North Adams employees are now learning painfully, service-type jobs in entities such as the Greylock Glen Resort or a proposed museum on the site of the old Sprague facilities offer neither the same level of pay or benefits as the industrial sector—if those service jobs ever materialize!

As you might suspect, my experience with the 1970 strike has left me with some strong feelings toward unions. Like most managers, if given the choice I wouldn't have a union, even one similar to the early independent ones at Sprague. While other people will disagree with me, I do not think they are required to give the American worker a fair deal. A successful and responsible company should do that anyway. Making employees into owners through ESOPs (employee stock option programs) can also bond the company and its workers closer together—that is, as long as the company continues to prosper. It is also unrealistic to guarantee "lifetime employment" forever. As all too many U.S. corporations have found, including leaders in electronics such as IBM and DEC who have tried to practice this, the business environment can reach a point where permanent layoffs are impossible to avoid. I also believe that the time will come before the end of this decade—if not sooner—that major corporations in Japan will be forced to violate this sacred pillar of their industry.

But unions are a fact of life. What does a company do if a union already exists or is successful in gaining representation? Simply put, management and union leadership *must* learn to work together so that the best interests of both workers and the corporation as a whole are served. Admittedly, this is not easy. But it can be done successfully. In the case of Sprague Electric, this was true from the late 1930s until the unfortunate work stoppage of 1970. And there are certainly examples where corporate and union leadership have worked closely together to create outstanding success. The October 21, 1991 issue of *Industry Week* carried its annual salute to "America's Best Plants." Included among the 10 plants named were two where close cooperation between management and unions was a critical factor in success. The first was the Cadillac Motor Car Division in Hamtramck, Michigan, whose workers were represented by the United Automobile Workers and which was a winner of the Malcolm Baldrige National Quality Award. The second was Johnson & Johnson Medical, Inc., in Sherman, Texas, whose workers belong to Local 513 of the United Textile Union. So despite my personal experience—and the preference of essentially all CEOs I know—American management and unions can work together successfully. The Japanese electronics industry has long been a good example of the benefits of such cooperation.

CHAPTER 7

Total Company Quality

A major part of training within Japanese corporations relates to the importance of quality. But, in general, the concept of quality goes far beyond the defect rate of products shipped to customers. Most (but not all) Japanese firms embrace the concept of *total quality control* (TQC). TQC—also known as *total quality management* (TQM) or *company-wide quality control* (CWQC)—has almost metaphysical or religious connotations to the Japanese. It involves all employees in the enterprise, from the lowest janitor to the CEO, in a complete, coordinated effort to continually improve performance at every level. As we have previously discussed, this thrust evolved from U.S. initiatives begun shortly after the end of World War II. But, as we will see in this chapter, the Japanese have added their own innovations—many of them deeply rooted in their national culture—to the U.S. initiatives. The result has been not only a high level of quality in their own products, but also the ability to influence the quality levels of some key U.S. electronic products such as DRAMs.

Learning From America: Japan and TQC

In early 1950, the Union of Japanese Scientists and Engineers (JUSE) started a publication titled *Statistical Process Control*. That same year, Dr. W.E. Deming was invited to Japan to give a series of seminars on the use of statistical process control (SPC) in manufacturing. One year later, using funds generated by Dr. Deming's lectures and consulting, JUSE sponsored creation of the Deming Prize. Because Japanese corporations are so fiercely competitive, this award served as an extremely effective form of competitive benchmarking and independent quality audit across industry and accelerated the drive for superior performance throughout Japan. Today, there are three different Deming awards, one each for individuals, companies, and factories, now entirely funded by JUSE. Teachings from the United States continued when, in 1954, J. M. Juran introduced the concept of quality control management. Even the suggestion system, a key part of TQC, was introduced by *Training Within Industries* (TWI) and the U.S. Air Force.

As time went by, the Japanese began to put their own stamp on quality. Unlike the United States, the importance of statistical techniques began to have a major impact on educational curricula. In 1962, the first *quality control circle* (QCC) was formed in Japan and, in 1966, Toyota received the first Japan Quality Control Prize, a highly prestigious extension of the Deming Prize. The Deming cycle, which preaches the continuous interrelationship between research, design, production, and sales to create superior quality and performance, evolved to the PDCA cycle. This is an acronym for *plan* (involving management), *do* (involving workers), *check* (involving supervisors), and *action* (management again). Japanese industry was off to the quality races, and TQC became the total embodiment of all these concepts.

But before we expand on how TQC is used in Japan, we need to determine just what we mean by "quality." There are as many definitions of "quality" as there are people to create them. According to Imai in his book *Kaizen*, quality relates to people. In other words, when given the right environment, education, and training, people will create high quality products with competitive performance and cost. As recently suggested by an associate of mine, this infers the definition that "quality is doing things right the first time." Referring to product quality, I like a less esoteric definition: "Quality is the meeting and maintenance of agreed upon performance specifications during the life of the product," Or, more broadly: "Quality is meeting or exceeding your customers' expectations." This last definition incorporates the concept of internal "customers" within a corporate entity and makes the all-important point that, no matter what a supplier may believe or say, it is the customer who has the final word on how good he or she is. The Japanese have incorporated the importance of the customer into a concept called *quality function deployment* (QFD), which was developed by Professor Yoji Akao in the 1970s. In this concept, activities which are unrelated to customer satisfaction are considered wasteful.

TQC within a Japanese company has many aspects, including the following:

- Total commitment and involvement by corporate management, especially the CEO. This embraces the continuous setting and monitoring of ever higher performance standards.
- Training at every level of the corporation including, in some cases, the board of directors.

- The design of product to simultaneously incorporate competitive performance, quality, and cost. This is often accomplished through use of cross-functional project teams. In electronic components such as IC memories, conservative designs and derating are believed to be major reasons for superior Japanese quality.
- Very close relationships with first-tier suppliers and subcontractors to ensure the on-time receipt of high quality materials, products, and services. This can include the requirement that they have their own TQC programs. The trend is away from the use of inspection, except as an audit function.
- A quality champion, who may be the head of quality for the corporation or even the CEO.
- Detailed involvement by operators through group activities such as quality control circles. An effective suggestion system is important as well.
- Continuous investment in capital equipment to replace direct labor. While productivity is an important goal here, more important is higher quality due to the elimination of human error. A properly designed piece of equipment will repeat the same operation over and over again with an accuracy impossible for a human being, no matter how well trained. Such equipment investment also requires strong maintenance programs such as *total productive maintenance* (TPM).
- The heavy use of statistical tools to, for example, control processes throughout the manufacturing system. Statistics are also used to eliminate waste, solve problems, and in the design of products, processes, and equipment.
- Competitive benchmarking through analysis of competitive product and programs targeted at industry awards such as the Deming Prizes.
- Detailed failure analysis capability. With parts-per-million (PPM) requirements now in the single digits, such analysis is the only way to determine the cause of failure and to develop solutions.
- Comprehensive and continuous information flow concerning performance to every level of the corporation. Such flow tends to eliminate the need for slogans which, I feel, are used all too often in the United States. Performance speaks for itself.

- Constant customer feedback to determine that expectations are being met or exceeded.

While these initiatives appear to relate primarily to product quality, it must be remembered that TQC covers all corporate endeavors, including the interrelationships of people, safety, cost reduction programs, meeting production quotas and deliveries, research and development, and responsiveness to customer requirements. A perfect TQC program would mean a perfect company.

How Sprague Exploited a Quality Advantage

Many people in this country believe that the perception of poor quality by U.S. manufacturers is skewed by the view we have of the auto industry and, in electronic components, by the quality of IC memories. Problems with the latter first came to light when, in 1980, the Data Systems Division of Hewlett-Packard reported that imported 16K DRAMs from Japan had significantly lower failure rates than those supplied by U.S. manufacturers. However, this unfortunate impression does a disservice to a number of U.S. component suppliers to whom quality and reliability have been a competitive advantage since long before World War II.

Using technology licensed from Sam Ruben, Sprague introduced in the mid-1930s a new "dry" electrolytic aluminum capacitor which had superior performance to the then prevalent "wets." "Dry" refers to the use of a non-aqueous electrolyte. In addition, modified packaging technology produced a radio power supply capacitor that kept the electrolyte from leaking, which was a major problem with competitive units and cause of radio failures over time. On a sales call to Philco, a silent "R. C." Sprague listened to the sales pitches of other suppliers as they all sat in front of the desk of the buyer, their capacitors lined up on his blotter. Finally, the buyer said, "Well, Mr. Sprague, you have been pretty quiet. What do you have to say?" Picking-up the Sprague unit and noting that the blotter was dry, "R.C." replied, "Ours don't leak!" Around every other unit there was a gradually spreading ring of electrolyte. This led to the Sprague slogan, "Second to None in Quality" which, in 1962, became the best known Sprague logo, "Sprague—The Mark of Reliability."

During World War II, Sprague had an unparalleled record of supplying electronic components of superior quality to the military services and received five Army/Navy "E" Awards and a Navy "E" Award for excellence, a record unequaled by any other component supplier. This later led to the designation of such components as "high reliability" and HYREL as a Sprague trademark.

Perhaps most illustrative is Sprague Electric's early experience with IBM. Sprague supplied IBM the first true high reliability component, a special film capacitor called the 195P. Because of the success of this unit, IBM then asked if Sprague could supply an extended range aluminum electrolytic which could operate out to 10,000 hours, far beyond anything that then existed in the industry. Sprague agreed to try, although they would not guarantee such performance since tests had never been run out this far. All qualification testing was completed successfully, including a 2000 hour life test. However, because of the extended life requirement, Sprague continued the test out to 10,000 hours. Around 7000 hours, occasional random failures began to appear which, after analysis, indicated that there was a highly infrequent aging problem which only occurred long into the operating life of the product. "R.C." personally called his senior IBM contact, Arthur Watson, the brother of IBM founder Tom Watson, to report the problem. After extensive further testing and consideration, IBM decided to retrofit all equipment that contained the Sprague units and to scrap all unused inventory at considerable expense to both corporations. In the meantime, Sprague diagnosed the source of the problem as the starting aluminum foil, switched to a higher purity material, and, some months later, was able to resume shipments. Because of the integrity and responsiveness demonstrated by Sprague's actions, IBM was to become one of Sprague's largest and most important customers in the years that followed despite the trauma of this unfortunate incident. This position was sealed when Sprague agreed to commit 100% of the first year's production of the new solid tantalum 150D capacitor to IBM.

These highly personal anecdotes—which I admit to being proud to relate!—show that, long before concern with product quality and reliability became fashionable, a young, aggressive U.S. company seized on this opportunity to create competitive advantage. In no small part this was due to its strong position in materials and processing technology. Even with concerted effort by competitors worldwide, this enviable advantage has lasted during much of Sprague's life. Although the designation didn't exist then, and SPC and quality circles were to come years later, there was much of TQC in the Sprague approach. This meant total Sprague commitment with the CEO as champion, operator training, conservative product design and derating, close ties with key suppliers, a very effective suggestion system, and competitive benchmarking through failure analysis of Sprague and rival products. It is also true that product integrity and reliability took precedence over delivery promises and, in general, Sprague was able to charge higher prices for its products.

Did the poor impression Sanken had of Sprague IC quality in the late 1970s mean that the decentralized semiconductor division didn't follow the basic corporate philosophy on the importance of quality? No, we embraced it fully. Most of our U.S. customers also bought the other Sprague component families, were leaders in their market segments, and wouldn't have allowed us to do so. For example, during our annual quality audits by IBM, our performance was compared against both other semiconductor suppliers and the rest of Sprague. In general, the latter comparison was the most demanding. On the other hand, the Sanken problems related to visual defects which—as far as we could determine—were of a very minor nature and had no impact as far as product reliability was concerned. The Sprague emphasis was on supplying reliable products that did not fail. In addition, the defects that Sanken found unacceptable were not rejects with our U.S. customers. Nevertheless, the Sanken experience did teach two very important lessons which are true for any American company that wants to sell its products in Japan:

1) What is acceptable in the United States may not be acceptable in Japan.

2) To the Japanese, such defects are a sign of an inability to meet an agreed-upon specification or commitment and—even more important—poor workmanship.

The United States and Quality Competitiveness

Regardless of the past, does the United States today really have an overall competitive quality problem in its manufacturing industries? While it is hard to generalize, let's discuss a few examples.

A 1991 article in *Ceramic Bulletin* estimated that poor quality within U.S. industry represents a cost of more than 25% of each sales dollar. While the semiconductor industry has closed the gap with Japanese suppliers in most product areas, quality is a moving target with ever tightening standards. Despite the recent financial problems of leaders such as IBM and DEC, the computer and telecommunications sectors appear competitive from both a quality and performance standpoint. (However, they also are major users of imported electronic components, primarily from Japan.) After slipping compared to both domestic and Japanese competitors during the 1970s and early 1980s, Xerox has been able to recover much of its position through a concerted, company-wide program that includes a major emphasis on product quality.

Unfortunately, the consumer electronics industry has moved almost entirely to Asia. Imported products, as well as those manufactured in the U.S. by foreign-owned subsidiaries, are generally of very high quality. With electronic components other than semiconductors, one needs to look on a company-by-company basis. In addition, more and more firms are being bought by non-U.S. corporations, as in the cases of MENA and AVX.

The truth is that standards are now so high on a global scale that quality has become only a competitive disadvantage. If your quality is not as good as the best supplier, you won't stay in business very long.

In addition to early and continuing emphasis on performance and reliability by companies such as Sprague, what are U.S. corporations doing to become more competitive in the field of quality? Before providing some specific examples, we should spend a few minutes discussing the two current major quality programs in the United States and in Western Europe. These are the international ISO-9000 series quality system and the U.S. Malcolm Baldrige National Quality (MBNQ) Award. While the latter is where we read all the media hype, the ISO system is quietly becoming a required standard both in Europe and the United States, where it is replacing the current military quality specifications.

The overseeing agency for ISO-9000 is the Geneva-based International Standards Corporation. The system has been adopted by the European Community (EC) Council, and quality professionals feel it will be required by most major customers in the U.S. and the EC by 1997. Like the current U.S. military standards, it is internally oriented and covers the complete quality system in a company. In addition to defining the responsibilities of management relative to quality, the system covers such important functions as contract administration, design, documentation, purchasing, process control, inspection and test, failure analysis and related corrective action, packing and shipping, the internal audit function, use of statistical techniques, training, and customer service. And, like military standards, the system requires meeting a defined quality level and, depending on the customer, may require independent third party certification by an agency such as Underwriter Laboratories (UL). Unlike the Malcolm Baldrige Award, there is little required in the way of customer feedback, although this will be addressed in future updates.

The MBNQ Award was created in 1987 by Public Law 100-107 to "help stimulate American companies to improve quality and productivity for the pride of recognition while obtaining a competitive edge through increased profits." Its model was the Deming Awards in Japan, and it is much less a quality system than a criteria-based contest. Unlike the ISO system, it doesn't measure quality as much as it does improvement in specific, quality-related categories. It is also heavily directed toward customer satisfaction. Each year there is the possibility of a maximum of two awards in each of three categories: manufacturing, services, and small businesses. While the Malcolm Baldrige Award has certainly sharpened our focus on the importance of quality, critics cite important negatives. Because of its emphasis on improvement, it doesn't necessarily reward the highest quality. For example, the much touted 1992 Cadillac Seville was built by the 1990 Baldrige award winner Cadillac Motor Car Company. However, it is still rated by experts as inferior to imports such as BMW, Mercedes, Lexus, and Acura. Marketing of the award often is misleading. As an example, a headline in the January, 1991 issue of The Institute reads "IBM and GM win manufacturing awards." True, but you have to read on to learn that it was plants or divisions, not the entire company, that was awarded. It is the same with other publications that praise the award. Winners have to share their experience and procedures with other companies, a process that can become extraordinarily time consuming. Applying to win the award is expensive, and, despite its lofty goals, the Malcolm Baldrige Award doesn't really address productivity. Perhaps most importantly, quality improvement is a never ending, long-term commitment which the Japanese have been pursuing for many more years than almost all U.S. corporations. Applying for, and even winning the Malcolm Baldrige Award, is only one step on a long and tortuous path with the Japanese still far in the lead in most industries. I hope U.S. corporations have the staying power. Table 7-1 summarizes the more important comparisons between the ISO and Baldrige quality initiatives.

TABLE 7-1

Comparison of ISO-9000 and MBNQ award

ISO-9000	*MBNQ Award*
International quality standard	U.S. quality award
Defined quality levels	Awards continuous improvement
Internal orientation	External orientation
Documentation oriented	Customer satisfaction oriented
Replaces U.S. military quality systems	Not a military quality replacement
Independent third party certification	No third party certification
Mandatory in three to five years	Some U.S. customers (like Motorola) require application
Cost for certification approximately $50,000 plus maintenance	Cost of application between $20,000 to $40,000 per plant
Improves quality	Improves quality

All other things considered, the last item in Table 7-1 is the most important. That is, both these initiatives do improve quality in a plant, division, or company. With the exception of the first item, which shows how the original pupil, Japan, has largely become the teacher. However, there are some more specific examples of quality programs within different U.S.-based corporations. Let's look at some of them.

At the Eastern Quality Conference and Exposition, sponsored by the Tower Conference Management Co. in Washington D.C. from February 5 to 8, 1990, there were Kaizen and Taguchi (a system of statistical manufacturing techniques developed in Japan) workshops. Lest we forget the original sources, there were also workshops by the Juran Institute and Philip Crosby Associates, Inc. (Quality is great stuff for consultants!) The Business Products and Systems Division of Xerox won the Malcolm Baldrige National Quality Award in 1989 as the result of what could only be described as a TQC program. And, incorporating techniques learned from their Japanese parent, MENA has established its own "1.0 QRS" Program as its method for qualifying for the Baldrige Award.

One of the most passionately vocal companies on quality and TQC has been Motorola, which found religion in the early 1980s as the result of the vision of Chairman Bob Galvin. Stung by their loss of position to the Japanese in television and auto radios, Galvin decided to push quality as a key strategy in order to remain a leader in wireless communications. In 1981, he set the lofty goal of improving Motorola's quality in product and services by ten times by 1986, set even more stringent goals in 1986, and created the "Six-Sigma" target for 1992. Six-Sigma is a statistical measure which is nearly equivalent to zero defects. More specifically, it allows 3.4 defects per 1,000,000 units or "events." To be a fully Six-Sigma company requires not only essentially defect-free products, but—and this is much harder to measure—perfection throughout the corporation. Motorola was also extremely clever not to use the TQC terminology, which has a Japanese connotation, but to put their own identity on the program. Whatever the name, the initiative worked. Motorola received the first Malcolm Baldrige Award in 1988, is the world leader in cellular telephones, and a major force in pagers and two-way mobile radio. They are also forcing the issue with their suppliers, requiring them to have formal Baldrige Award programs if they wish to continue to supply to Motorola. While hindsight is easy, the world situation in consumer electronics might have been very different if the company had applied the same zeal to its television and radio businesses many years ago.

Following the Motorola lead, IBM has created its own Six-Sigma program. In 1990, the Rochester, Minnesota plant received the Baldrige

Award for its AS/400 midrange computer. Similar agendas are underway at Corning, Northern Telecom, Boeing, Caterpillar, Digital Equipment, Raytheon, and elsewhere. Allegro Microsystems has recently instituted its own TQC program which leads its Japanese parent, Sanken Electric. So far, Sanken has concentrated on quality circles and is monitoring progress at its Worcester-based subsidiary.

The Sprague Electric Solid Tantalum Capacitor Division has a total quality program that encompasses product and materials development, process control, training, PPM standards on outgoing quality, customer partnerships, and ship-to-stock programs. Each capacitor shipment includes a statistical process control/capability report that charts overall process control during the period when the lot was manufactured.

Our discussion so far makes an inescapable point: because of global competition, survival in electronics will be very difficult—if not impossible—without a corporate-wide total commitment to quality, whether "TQC," "TQM," "Six-Sigma," or whatever. Such an effort must be led by the CEO and embraced by every employee.

Moreover, materials processing industries, such as semiconductors and electronic ceramics, require a special emphasis on, and knowledge of, the relationships between the nature of the starting materials, their processing, and the quality and performance of the finished devices. While this may sound self-evident, several additional examples from my experience will help underscore this critical point:

- While assessing quality problems and major yield variations in a captive ceramic component operation, I was amazed to learn that the cheapest of all possible starting ceramic powders was being used without any guarantees by the supplier or even incoming evaluation by the user. The ceramic powder source was also notorious for variations in the supplied materials. In addition, the process included a firing step which allowed a variation of no more than +/- 15° Celsius. Unfortunately, the old kiln involved operated over a range of +/- 50° C. Because capital was tight, replacing the kiln was not an option. However, working with an outside laboratory, they were able to qualify a modified formulation that was not only guaranteed in relation to parameters such as impurity levels and particle size and distribution, but which also would provide the required performance over the full profile of the kiln. As a result, the problems largely disappeared. How could such a situation arise in the first place? The management of the plant just plain had inadequate experience in ceramic technology to understand the implications of actions they had taken, primarily to reduce costs.

- Some idea of the importance of the starting material can be deduced from Figure 7-1. At high magnification, it shows the difference between the uniformity and topology of two types of barium titanate, one prepared using conventional solid state processing while the other is chemically synthesized fine powder. The uniformity of the latter has been found to yield important benefits. First, in the fabrication of MLC ceramic capacitors with the same dielectric thickness, the fine powders give both higher electrical performance and reliability. Second, such layers can be made much thinner thus yielding components with considerably greater volumetric efficiency. The much higher cost of the chemically synthesized material is more than justified by these advantages.

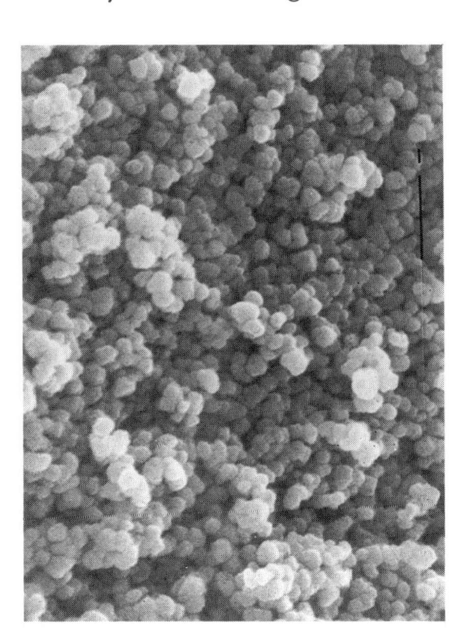

Figure 7-1: Differences in the uniformity and topology of two types of barium titanate. The chemically synthesized powder is finer and more uniform

- On another occasion I was asked to recommend a ceramic substrate to be used for a new type of deposited thin film resistor. While cost was important, quality was even more so. As the result of literature research and personal interviews I was able to prepare Table 7-2 which compares the characteristics of different alumina substrates as a function of the purity of the starting material. There are some interesting trends. For example, density, bend strength, and thermal conductivity all improve with purity. For the particular application of interest, surface roughness was most important, and was also shown to improve with purity. Therefore, a 99.5%

pure "thin film substrate" was chosen, despite the much higher cost compared with less pure material. Use of aluminum nitride (AlN), with its much higher thermal conductivity, was also eliminated because of reported problems with the reliability of deposited thin film resistors caused by the greater surface roughness of AlN substrates.

TABLE 7-2

Properties of alumina substrates as a function of purity

	Thick Film	*Thin Film*	*High Purity*	*Single Crystal*
Purity (%)	96.0	99.5	99.7	99.9
Cost ($/lb)	2			38
Surface roughness	0.3	0.075	0.04	0.04
Density (g/cm^3)	3.7	3.91	3.95	3.97
Bend strength (MPa)	330	400	300	700
TCE (10^{-6}/C)	7.7	8.1	8.1	6.8
Thermal conductivity (W/m.K)	20	30	30	40
Dielectric constant	9.3	9.7	9.9	9.2
Dielectric loss (10^{-4})	3	<1	<1	<1
V_{Br} (KV/mm)	>10	>10	>10	>10
Resistivity (Ω/cm)	>10^{14}	>10^{14}	>10^{14}	>10^{14}

- Figure 7-2 is a "Quality Flow Chart" for the front-end, or chip manufacturing, process for an MLC capacitor. This happens to be one for Sprague Electric, but it would be similar for other manufacturers. This flow chart is interesting because it not only shows the number of processes involved, it also details the control characteristics that must be measured and managed and the related quality control documents. As far as this discussion is concerned, we see that, in addition to the starting ceramic materials, we must also be concerned with electrode pastes, termination systems, plating solutions, and appropriate solders. As already discussed in Chapter 3, there is also a great deal to be mastered to get the starting ceramic powders into usable form. We will have more to say about the impact of starting materials in the next section which deals with a subject that, for some reason, is strikingly absent in most quality discussions.

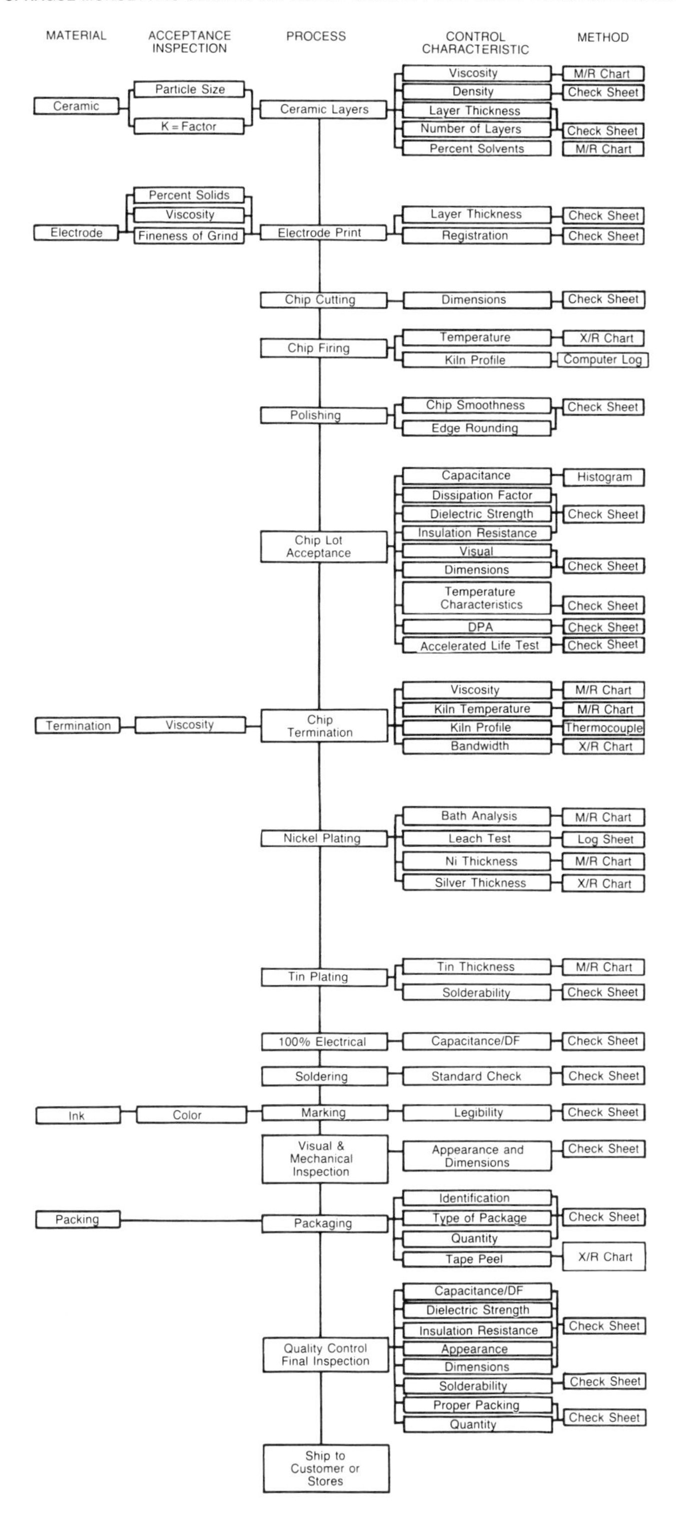

Figure 7-2: Quality control flow chart for Sprague capacitor manufacturing.

None of the examples above is as dramatic as a major company-wide quality campaign announced with trumpets. But they do illustrate how quality can be improved in significant ways through small, easily overlooked steps. This is the way the Japanese have done it for years, and it is also the way the United States must start doing things.

Quality and Reliability

If you look for the term "reliability" in the indexes of any of the books I have cited so far, you will not find it. However, you will find "quality." Many people tend to equate quality and reliability. My definition of reliability is the absence of defects resulting from storage or use conditions. Reliability is certainly inferred in the concept of meeting customer requirements (who wants to receive goods that fail under use?), but the emphasis by Japanese firms appears to be more on the elimination of defects, conservative designs, and derating than on the detailed understanding of the relationship of defects to the physics of failure. If one uses conservative designs, optimum materials and processes, and eliminates defects, then one might assume that long life under the use or storage conditions should automatically follow. However, quality does not necessarily take into consideration aging phenomena, which can occur due to changes in material properties over time and under use. For example, an initially defect-free car may develop problems due to rusting-out caused by inadequate undercoating and the use of thin body steel. Some early Japanese models had such a problem. Internal components such as water or fuel pumps may fail over time, or electronic equipment may develop problems due to the hostile automotive environment. Detailed analysis of such failures leads to improvement in later models. Frequency of repair data for Japanese autos indicate that this lesson has been learned well.

Since its inception, the IC industry has had its share of reliability problems that were solved only after the fact. Probably most well known was the "purple plague," a brittle, poorly conducting intermetallic phase that led to lead attachment failure in semiconductor devices that were improperly assembled or stored at too high a temperature. The introduction into MOS devices of alkali metal impurities, such as sodium, can lead to changes in the gate surface charge over time and subsequent device failures. More recently, "soft errors" have been caused in MOS ICs by alpha particles—that is, helium nuclei emitted from the sun's nuclear reactions. Small amounts have also been found in certain ceramic packages. This is one of the areas where Japanese 16K DRAMS were found superior to their U.S. counterparts due to more conservative designs, development of better passivation schemes, and superior ceramic packages. Other, yet-to-be determined phenomena should be anticipated as device geometries continue to shrink.

The device failures in the extended-range aluminum capacitors made by Sprague for IBM (discussed earlier) were due to minute levels of impurities in the foil and found only after thousands of hours of use. Higher purity foil solved the problem. In another case, dendritic growth of silver has led to failures in thick film resistor networks. Such growth is caused by the creation of an electrolytic cell due to the penetration of moisture into the package over time.

Reliability testing of MLC capacitors goes back to the 1960s, much of it under the aegis of the U.S. government. The goals of such work have been to find out and, if possible, eliminate the causes of various types of failures and to develop accelerated testing and screening methods. A variety of different types of failures have been found, some due to the nature of the materials used (for example, electronic disorders), and some caused by processing (such as cracking, delamination, and porosity). Such studies are critical since, for example, "wet" processes used in chip manufacturing have been found to reduce failures caused by delamination.

The Japanese approach is driven by elimination of defects, and very rapid turn-around in analysis of failures so that corrective action can be taken. Recalling our earlier discussions about Sanken, we had considerable difficulty understanding why they were so concerned with visual and mechanical defects. Certainly a bent or nicked package lead can cause a real problem. On the other hand, there should be minimal hardship caused by a smudged marking unless, of course, it could cause misuse. However, to the Japanese *such defects mean poor workmanship and that is not acceptable.* To many American companies, such Japanese complaints seem like nit-picking; to the Japanese, they are key indicators of the ultimate quality of the product.

Back in the mid-1980s I visited a TDK MLC factory in Japan and noted that, on one side of the factory floor, there were virtually no people to be seen manning the chip-manufacturing equipment because of the high degree of mechanization. In contrast, there were several hundred workers at microscopes on the opposite side looking for visual defects because no equipment had yet been developed which could meet the incredibly tight requirements. I would bet that today both sides are empty!

Experts such as Galeb Maher (president, MRA Labs) and Ken Manchester (an industry consultant) support the concept that U.S. firms, in no small part driven by government requirements, have been much more thorough than their Japanese counterparts in relating device failures to the physics of failure. Manchester, in particular, expressed surprise that his Japanese contacts had done so little in the way of long term reliability test programs and accelerated testing when compared to U.S. companies. On the other hand, the stringent demands of Nippon Telegraph and Telephone (NTT) were one of the major causes of the more conservative designs and packaging approaches used by Japanese DRAM manufacturers.

These observations relative to reliability were supported during a visit I made while writing this book to the Computer-Aided Life Cycle Engineering (CALCE) Center at the University of Maryland in College Park. The center is an existing National Science Foundation/Industry-University Cooperative Research Center (NSF/IUCRC). Building on its pioneering work in highly reliable military packages, it has undertaken a new initiative in development of commercial semiconductor packages. The center director, Professor Michael Pecht, argues that the United States has a real lead over Japan in the area of reliability science and understanding, at least in semiconductor packaging. He states further that the current techniques for reliability estimation and design, as required by the U.S. government, are inadequate since they do not comprehend wear mechanisms. "Rather," Professor Pecht told me, "reliability goals should be related to lifetimes and based on a detailed understanding of physical phenomena as they relate to the physics of failure." It is interesting to note that, here again, the U.S. "lead" is based on a deeper understanding of basic science. It remains to be seen whether this can be translated into superior U.S. products.

TABLE 7-3

Comparison of total quality programs in Japan and the United States

Element	*Japan*	*United States*
Designation	TQC/TQM/CWQC	TQC/Six-Sigma
Quality	relates to people	relates to product
Methodology	continuous improvement	quantum leaps
Reliability	defect prevention	physics of failure
Champion	can be CEO	usually QC manager
Training	all employees	primarily operators
Group activities	quality control circles	quality control circles
Suggestion systems	most companies	some companies
Statistics	heavy use	increasing use
Design	conservative; derating	push limits
Failure analysis	very responsive	responsive
Use of capital	heavy investment	moderate investment

Table 7-3 summarizes the differences in Total Quality programs at medium-to-large Japanese corporations with those at "superior" U.S. companies. This latter designation is difficult, if not impossible, to define. However, a number of such firms have already been mentioned in our prior text.

Relations with Suppliers

Most references cited previously in this book pinpoint close relationships with key suppliers as a Japanese strength and a U.S. weakness. For example, in his book *Kaizen* Imai identifies supplier relations as one of the principle elements in the TQC system as practiced in Japan. The MIT Commission's *Made in America* describes such relations as often being weak in the United States. In *Competitive Advantage*, Michael Porter reports that the clustering of related and support industries is an essential part of creating competitive advantage.

Additional sources beyond those already noted make the same point. In his 1991 scholarly treatise on the DRAM industry, *Technological Competition in Global Industries*, David Methe' goes further and points out that such relationships also spur the diffusion of technology. To the man—remember, this isn't sexist usage when discussing the Japanese electronics industry!—my Japanese contacts point to the importance of suppliers. (However, they also admit how hard it is for a foreign supplier to become one in Japan, especially at the first tier level.)

What is going on here? Why is this even an issue? One good reason is the topic of this chapter. The relationship of Japanese companies to their suppliers, both of raw materials and capital equipment, is a major reason for the quality of Japanese products. If a company gets higher quality raw materials and the latest in production and test equipment, it is obviously easier for that company to produce a high quality product.

In materials processing industries such as semiconductors and other electronic components, it is difficult—without close and ready access to your supplier base—to be sure of receiving the latest in processing equipment or stay current with the latest technology in, say, defect-free silicon wafers, electronic ceramics, or fine-powder tantalum material for higher CV capacitors. Major U.S. electronic OEMs (original equipment manufacturers) learned this years ago. In order to stay current with leading edge technologies, the top electronic OEMs worked closely with the top component suppliers, often jointly developing a particular component, or even component family, that provided them a unique competitive advantage. On their end, the component suppliers established close ties with their first tier materials suppliers, often with cooperative development activities to create the next generation component. As specific examples in the capacitor field, one can point to development of aluminum foil with a higher etch ratio, and use of higher CV tantalum powder to yield components with greater storage capacity. This approach is one of the ways Sprague Electric built its relationships with firms like IBM, AT&T, Delco Electronics, and Hewlett Packard. Sprague targeted the leading OEMs in each market segment, not only because they represented the largest business potential, but also because "specials" for them usually became industry standards over time. Such linkages are becoming even more important today as customers move toward "preferred vendor" programs and reduced numbers of suppliers.

Yet, having said this, something is obviously wrong. The independent United States passive component supplier is disappearing and the overall U.S. merchant semiconductor industry is clearly in trouble. Part of the latter problem relates to a rapid deterioration of the equipment and materials infrastructure in the United States necessary to support the industry and allow it to produce quality products. There is no lack of evidence to support this observation. In the February 28, 1990 issue of *Electronic Buyer News* (EBN), I wrote an opinion piece titled "Why Should We Care?" In it, I pointed out that more than 95% of the silicon wafers are now sourced by either Japanese or European owned corporations and that more than 90% of the ceramic substrates and ceramic packages used for IC interconnect and packaging come from Japanese owned corporations. To prevent overdependence on Japan, Sematech—an Austin, Texas-based research consortium involving several companies and some government funding—has been forced to look to Europe for both semiconductor equipment and materials (assuming competitive sources of supply can be found). Acquisition of Union Carbide's polysilicon business by Komatsu, Sony's takeover of Materials Research Corporation, and Nippon Sanso's bid for Semi-Gas helped lead to this. The situation is no better in equipment areas such as lithography; for example, GCA continues to lose position to Japanese corporations such as Canon and Nikon. Sematech recently charged that Japanese firms are withholding leading-edge production equipment from U.S. manufacturers, although major U.S. semiconductor companies such as Intel, Texas Instruments, and Motorola did not support the charge.

With unusual apathy, the U.S. government, as represented by the inter-agency Committee on Foreign Investment in the United States (CFIUS), has done little to block the continuing buy-out of the bedrock U.S. semiconductor infrastructure. Moreover, it is not even very clear that the U.S. IC manufacturers really care. I suppose the good news is that such foreign investment has saved American jobs that would have disappeared if the firms had gone out of business. But why are the purchasers foreign and not U.S. companies? I just cannot buy the argument that it doesn't matter who owns these companies as long as U.S. jobs are saved. If the entire infrastructure that supports an industry comes under foreign ownership, especially if some of the owners are also direct competitors to the affected U.S. customers, it is hard to believe that the foreign investors will not make decisions that are in their own best interests. Some specific examples which will be given shortly support this concern.

The Japanese have a very clear strategy related to electronics. After World War II, they concentrated on the consumer industry. The first important move in the industrial area came in telecommunications, and then, in the 1970s, the computer industry was targeted. Recognizing that the enabling technology in this industry was integrated circuits, this area was targeted as well. The major thrust behind the 1975-1979 VLSI Research Cooperative, which involved NEC, Hitachi, Toshiba, Fujitsu, and Mitsubishi, was to make these firms competitive with IBM. In *Technological Competition in Global Industries*, Methe' argues that, as the end result of such targeting, it is no accident that today IBM is the third largest computer supplier in Japan while it is number one everywhere else in the world. Each of these Japanese competitors is also part of one or more of the powerful groupings of companies called *keiretsu* which grew out of the large industrial holding companies that were banned by the U.S. after World War II. Because of the importance of this unique form of Japanese capitalism, we need to spend a few minutes describing them in more detail.

Two recent articles provide excellent insight into the keiretsu system. In "Inside Japan Inc.: Cozy Ties Foster Political Friction," in the October 7, 1991 issue of *The Washington Post,* Paul Blustein characterizes the inter-company ties that exist within the keiretsu as the principal source of Japan's economic might. There are two basic forms. The horizontal keiretsu includes companies in different industries. An example is the Mitsui grouping, which includes Toyota and companies in such diverse other industries as retailing, textiles, steel, real estate, construction, food, paper, and chemicals. In a vertical grouping, such as the Toshiba keiretsu, the ties are between a manufacturer and its suppliers and distributors. The two overlap since Toshiba is also a member of the Mitsui group. But these groupings are much more powerful than merely a loose association of corporations which involves cross-ownership and other financial and personal relationships. As sources of ready capital, they also include one or more large financial institutions at the heart of the group. Once criticized by Western economists as too inefficient to survive, they maintain their vitality through the fierce competition that exists between corporations that are in the same industry but are members of different keiretsu.

The cover story in the September 24, 1990 issue of *Business Week* described the hugely successful Mitsubishi keiretsu. While hundreds of interrelated companies are involved, at the heart of the 28 core members are Mitsubishi Corporation, Mitsubishi Heavy Industries, and the Mitsubishi Bank. Ownership by other group members in these three behemoths is 32%, 20%, and 26% respectively. Within such a keiretsu, strong vertical connections exist between electronic OEMs and their semiconductor suppliers. This is one of the major reasons that U.S. semiconductor companies have such a difficult time selling to Japanese computer manufacturers. They already have their own sources of supply within the keiretsu. The same can be said of the automotive manufacturers and their auto parts suppliers. I do not feel it is "Japan bashing" to note that overcoming the barriers for American companies trying to sell in Japan created by these industrial groups is an overwhelming task. As an analogy, imagine Japan telling the United States it needs to change its entire economic system to make its markets more available. Even Japanese suppliers find it close to impossible to sell to a potential customer within a group to which they don't belong as long as there are adequate internal sources of supply. Where such do not exist, the task is still daunting.

While U.S. component suppliers often are closely linked with their customers through long term, usually non-equity relationships, linkage backward into the supplier chain is more tenuous. On the other hand, the Japanese believe in trying to control all parts of the supply chain, whether within or outside of their own group. It is obviously easier for a company to enforce a TQC program if it has control over all elements that go into a product. As part of this approach, first tier suppliers are a key part of the value-added chain.

Besides gaining access to state-of-the-art technology, the Japanese strategy of systematic buyout of the U.S. semiconductor infrastructure appears based on additional goals, such as to gain control of a key part of the manufacturing chain, to better serve overseas markets, to create the necessary local infrastructure as they build or buy overseas semiconductor facilities, to buy valuable assets and resources inexpensively, and to acquire what they feel are, or can be, good businesses. Interestingly enough, they have been practicing exactly the same strategy in the automotive industry. What are the potential negatives as far as the U.S. semiconductor industry is concerned? A few specific examples provide some frightening insight. In the mid-1980s, the Sprague semiconductor group had a military IC program that required purchase of at least

50% of the ceramic packages from a U.S. source, in this case General Electric. Because of continuing problems with the GE product, Sprague became increasingly dependent on the other supplier, Kyocera. Then Kyocera moved production to San Diego, and lost the recipe. No packages meant no ICs, and a critical U.S. military program was in jeopardy. Sprague had little luck getting the Japanese company to supply out of Japan because, they claimed, "that production is completely allocated to our Japanese customers." Only direct intervention by the U.S. government was successful in reinstating package supply. If it had been an industrial program, the answer would probably have been "too bad."

During a trip to Japan in June of 1991, one of my most reliable contacts told me that certain Japanese DRAM manufacturers were receiving more defect-free Si wafers from the Japanese suppliers than U.S. customers were. The resultant yield improvement was as high as 10% in some of the newer devices such as the 4 megabit (M) DRAM. 10%!! However, this was not because the wafer supplier wished to provide the Japanese IC firms with a competitive advantage. The same thing could have occurred between two different Japanese customers. Rather, the Japanese semiconductor companies were much more demanding than their U.S. equivalents, would only accept the lower defect levels, and worked more closely with them to achieve the required result. While there may be tight linkages between customer and supplier in the U.S., it is generally accepted that Japanese OEMs are more demanding of performance from their sources than is common in the U.S. This is another U.S. competitive disadvantage cited by Porter in *Competitive Advantage of Nations*. Dr. Denda's words clearly show who is the top dog between customer and supplier: "There is a big difference in the relationship with the supplier between the States and Japan. In this case we are the customer for the supplier. They will try to satisfy our requirement. Accordingly, we are not very friendly to them."

Earlier we cited the debate between Sematech and certain U.S. IC suppliers on whether or not U.S. firms were able to get the latest manufacturing equipment from Japan. I believe both sides are correct. Where the supplier serves the open market, the same equipment is sold around the world. What is not generally available, not just to U.S. IC houses but also within Japan itself, is proprietary equipment designed in-house by a particular semiconductor manufacturer, or subcontracted within the keiretsu. This is no different than what occurs in the U.S. when equipment is so developed. However, with the strong skills the Japanese have in equipment design—developed both internally and through acquisition of U.S. companies—there seems little doubt that the Japanese are developing a worldwide competitive advantage that only strengthens as the U.S. infrastructure disappears. In my opinion, this continuing trend seriously jeopardizes the long term viability of the U.S. semiconductor industry.

As we can see from the examples above, the evidence is clear: strong links with suppliers of both raw materials and capital equipment has a strong impact on quality. The Japanese practice of developing such links is in contrast to the often-adversarial relations between companies and their suppliers found in the United States.

Are the relationships with suppliers similar with other components such as capacitors? While there are similarities, there are also major differences. First of all, the capacitor industry is much less capital intensive than in ICs. This is seen in the make-up of the industry. While there are huge Japanese corporations involved—for example, Matsushita in aluminums and NEC in tantalums—some of the leading companies are independents without any apparent keiretsu-like relationships. These include Murata, the world leader in ceramic capacitors, and Nichicon, one of the leaders in the aluminum area and in tantalums. As such, these companies have had to succeed through their own wits without the back-up of the huge conglomerates. This is very similar to what remains of the merchant capacitor industry in the U.S.

In the area of raw material supply for the capacitor industry, the Japanese operate as they do with ICs, and have close but extremely demanding linkages with first tier suppliers. In ceramics, many, such as Murata, manufacture much of their own material. In the United States, material is more often purchased, usually from Japanese sources. Smaller Japanese firms or divisions, such as the ceramics division of Mitsubishi Materials, purchase on the open market with, however, very tight control of incoming material. In aluminum capacitors, most of the larger Japanese manufacturers etch and form their own foil, as was true some years ago in the United States. Today, the remaining U.S. firms increasingly purchase their foil around the world. Either approach can work, assuming there is tight coupling between customer and supplier. This becomes more difficult the further the geographic separation and favors the Japanese as corporations within their country increasingly become the world sources of both materials and equipment. The same can be said of the relationship between the end equipment manufacturer and component supplier. With the purchase of AVX by Kyocera, the ceramic capacitor business in now controlled by Japanese companies, a situation that should concern U.S. electronic OEMs. Since loss of the ceramic package business didn't, this probably won't either.

Most of the Japanese capacitor manufacturers have a deep, direct involvement in the design of manufacturing and test equipment. Some, such as Matsushita and Murata, manufacture most of their requirements internally. Murata even sells surface-mount attachment equipment to their customers. In the case of Nichicon/Sprague (N-S), while the conceptual work is done in-house, most of the manufacturing is subcontracted on the outside, usually to local machine shops. This is similarly true at Mitsubishi materials. The situation is similar with many successful firms in the United States, although there is a greater tendency to buy in the open market, again often from Japan. Problems experienced by Sprague Electric related to such purchases have been two-fold: first, the latest generation of equipment is seldom available, and, as received, the equipment never fully meets specification. This probably relates to the difference between proprietary and open-market designs, as discussed earlier with semiconductors.

The second complaint provides some interesting philosophical perspective. *Whether designed and manufactured in-house, subcontracted, or bought in the open market, the Japanese user automatically assumes that the equipment will not initially operate as originally specified.* This is not the result of incompetence on the part of the supplier. Rather it results from the fact that it is almost impossible to completely specify a new piece of manufacturing equipment until it has been in actual use for some time. Therefore, every factory has its own, internal equipment group that not only can maintain and repair the equipment properly but can also either make simple modifications directly or work with the original source on more substantial ones. This is a capability I have seen lacking all too often in U.S. capacitor operations. Jack Driscoll of Murata Erie North America (MENA) provided me further insight from his company's perspective. Murata never commits a new piece of production equipment to the manufacturing floor until two conditions exist: there is complete understanding concerning how the materials and processes involved relate to the equipment; and the next generation replacement equipment is on the drawing board. This is another example of the Japanese approach of continual improvement and upgrade of standards.

Peter Maden feels that, at least in solid tantalum capacitors, the situation is little different between the U.S. and Japan and, in fact, favors the former (in other words, *his* tantalum business). For example, for many years, because of the economies of high volume purchasing, he has bought and supplied all the tantalum raw material for both the Sprague Electric manufacturing locations and the Japanese subsidiary, Nichicon-Sprague. Both Sprague and N-S compete very effectively with the other worldwide competitors. He also claims that, if one is willing to work hard enough at it, it is possible to purchase state-of-the-art manufacturing equipment directly from Japan. I think the key words are "hard enough"!

I feel the competitive situation in capacitors is less desperate than it is in semiconductors. However, there are many similarities and the trend is the same: the independent, U.S.-owned and headquartered capacitor manufacturer is slowly disappearing. In my opinion, the U.S. electronics industry will suffer over the long term because of such trends.

Supplier Links, Quality, and the Shell Club

I would like to close this section on a somewhat lighter note that ties together some of the points discussed in this chapter. While extremely demanding of their suppliers, Japanese companies can practice some non-traditional methods of creating linkages with both their customers and suppliers. I became involved in one during periodic visits to Kyoto as a director of Nichicon-Sprague. Sprague Electric was much more important to the venture than just as an investor; it also served as both a customer (of N-S capacitors to be sold in Europe and in the U.S.) and a supplier (of technology and tantalum powder) to the joint venture. Therefore, bonding with the Sprague board members was obviously important.

While full of content, there was a repetitive and ritualistic aspect to each of our meetings. We were always met at the airport in Osaka or train station in Kyoto and taken directly to the hotel, generally the Kyoto Grand, and left alone for the evening. The next day, there was a visit to the factory (the best part of any board meeting), dinner "with the troops," and finally an evening on the town. The board meeting was held the next day in a smoke-filled room at the Nichicon Capacitor corporate headquarters in downtown Kyoto (as a non-smoker I was never sure I could last through the day). There was a more formal dinner that evening with N-S president Hirai, and again we headed for the bright lights. Wisely, Hirai-san never joined us. Usually our destination was the Shell Club, one of the hundred or so private business clubs that dot Kyoto and serve as the center of after hours business—if you can call it that! Many of the linkages between companies and suppliers that we have examined so far in this chapter have been forged in the environment of places like the Shell Club. Nichicon wanted a good working relationship with Sprague Electric, so I and the other members of the Sprague Electric team became regulars there.

"Business" at the Shell Club consisted of cognac or liquor, more cigarette smoke, dancing with bored girls who tried to make conversation in broken English interspersed with giggles, and the singing of karaoke songs with our hosts. I have no idea how many times I hoarsely crooned "I Left My Heart in San Francisco." Once my wife came along with me to such a club. As the only Caucasian woman I had ever seen in one, she was inundated with requests from perspiring, slightly intoxicated Japanese businessmen to both sing and dance. "Never again!" was her summary comment on the proceedings.

But to dismiss such evenings as adolescent male behavior would be wrong. Some important things were going on underneath the surface foolishness. During all this, Pete Maden and his N-S counterpart, chief engineer Mitsui-san, would huddle in the corner designing the next generation tantalum manufacturing equipment—in between drinks and dances, of course. Forever bonding, Ike Takeda of Nichicon usually was at my side, asking endless questions about the industry, competitors, Sprague Electric, and the like, most of which I either could not or would not answer. As this routine continued year after year, I began to politely bow out, preferring to brush up on my Japanese at the television in my room. The protestations of my hosts, especially Takeda-san, seemed only out of politeness; I was certain that everyone would be happier without my company. I only learned much later that this repetitive, exhausting, and increasingly boring ritual was an important part of the business relationship. Not only did the mutual silliness of our actions create unity—somewhat like school boys on a forbidden fling—but, as Ike told me recently, "useful information was also gained that could be learned almost no other way." (As the old Navy saying goes, "loose lips sink ships!")

Obviously, I am not advocating wild nights on the town as the solution to America's quality problems. But it is clear that the Japanese way of achieving quality involves much more than just an adherence to the principles of Deming and Juran. It also involves suppliers and their ability—or willingness—to share their customers' commitment to quality. Obviously, something exactly like a keiretsu system is highly unlikely to evolve in the United States (if for no other reason than potential anti-trust and related legal problems). However, the United States needs to develop some way of creating closer relationships between corporate customers and their suppliers. This is important not just for the sake of quality itself, but also to make sure we have the ability to control our own "quality destiny."

CHAPTER 8

The Manufacturing Floor

We have already "walked around" the manufacturing floor. In earlier chapters, we described in some detail how a multi-layer ceramic capacitor is made and detailed the use of cross-functional teams to design cost effective, quality products. We have compared the U.S. work force with that of the Japanese (with proper training, not too bad), talked at length about U.S. quality (improving, but chasing a moving target), and discussed relations with suppliers and the status of the United States electronics infrastructure (not so good). But now it is time for us to go out onto the manufacturing floor itself.

They speak a special language "on the floor." WCM (world class manufacturing) seems to consist of acronyms and strange sounding descriptors like JIT (just-in-time), CAD (computer-aided design), CAE (computer-aided engineering), CAM (computer-aided manufacturing), CIM (computer-integrated manufacturing), MCAE (mechanical computer-aided engineering), and TPM (total productive maintenance). Kamban (label) cards accompany material as it moves between manufacturing stations or *cells.* "Flexible manufacturing" machines are designed for quick changeover and use *poka-yoke* (fail safe) techniques such as guide-pins and limit switches. *Jidohka* or autonomated (yes, the spelling is correct) equipment automatically stops if a problem occurs. Perhaps the competitive problems the United States has on the manufacturing floor are due to using the wrong dictionary! But there are indeed problems for the United States when it comes to manufacturing. Perhaps in no other area does Japan hold such a commanding—and from all available evidence, widening—advantage.

The authors we've cited earlier agree with that assessment, although for varying reasons. In *Competitive Edge* , it is hypothesized that Japan's leadership position in the use of automatic production equipment may be due to the decision by U.S. semiconductor manufacturers some years ago to use cheap offshore labor rather than mechanize. The MIT Commission's *Made in America* says the U.S.'s weak position in manufacturing is due to poor designs from a manufacturing and quality standpoint, lack of attention to manufacturing, and decay in the competence of the United States machine tool industry relative to both Japan and West Germany. Porter in *Competitive Advantage* cites other reasons, such as the current dominance of the robotics industry by Japan (although the industry actually started in the United States in the 1950s). The growth of robotics in Japan was a necessity, caused by high growth and an increasingly limited and aging labor pool. (This last point may have some interesting long term consequences as the Japanese economy matures.)

The overwhelming use of capital equipment plus lifetime employment means that most costs in Japanese businesses are fixed. To some degree, this is compensated for by the use of part-time labor and subcontractors. Nevertheless, the overwhelming drive for market share and growth leads to what I call "the fixed cost cycle." Capital investment plus lifetime employment requires continuing growth and market share increase, which in turn requires further investment and people, and the cycle goes on and on. As long as growth continues, everything is fine. But, in the case of an extended downturn, such a system can lead to financial disaster for a company or an industry, as we have found out all too painfully in the United States components industry this last decade. This is one of the reasons that lifetime employment in Japan probably has a limited future. At the time this book was being written in mid-1991, there were signs this cycle was coming under increasing strain due to the impact of a worldwide recession on the Japanese economy and the effect of deflation on the Japanese real estate and stock markets.

What is World Class Manufacturing?

The phrase "world class manufacturing" is a popular buzzword in American industrial circles these days. But just what is meant by the term?

While there are many different views of what constitutes world class manufacturing, there is reasonably consistent agreement on the major components. Actually, *world class business* (WCB) might be a more appropriate term. We have already discussed a number of the important criteria in the earlier section on quality. The following are the key items of what I prefer to call a WCB. Those parts which are central to manufacturing itself are JIT, flexible manufacturing, the subdivision of the manufacturing flow into sub-manufacturing units or cells, and mechanization to replace labor.

World class manufacturing starts with involved and knowledgeable senior management. This means CEOs who not only understand their businesses from the ground up, but are also involved in the details of running the enterprise. This tends to be more true of Japanese management than in the United States where—all too often—CEOs spend an inordinate amount of time on financial analysis and relations with their board of directors and stockholders. Management must create a corporate strategy that is consistent with both the needs of customers and the resources available to fulfill these needs. While this might seem self-evident, understanding the corporation's available resources, both financial and in competitive technology, is critical to prevent strategic and tactical decisions that cannot be fulfilled. As just one example, a decision to undertake a needed but expensive mechanization program may fail due to the lack of adequate internal engineering and financial resources. Competitive benchmarking is a critical input.

World class manufacturing demands a global orientation. Since the world's markets and competition are increasingly worldwide, there are at least three reasons why such orientation is critical. For one, it creates greater market potential. A second is that competing in a local market against the top local suppliers is one of the best methods of competitive benchmarking. Finally, local presence is one of the best ways to gain access to local technology. However, such orientation must also be affordable. For example, penetration of the Japanese market can be an extremely expensive undertaking.

Several aspects of world class manufacturing are similar to those practiced in Japanese manufacturing, such as close links with both customers and suppliers. We have previously discussed this point, but it bears repeating. There is an increasing trend worldwide for electronic OEMs (original equipment manufacturers) to strictly limit the number of top tier suppliers and work very closely with them. This is especially so in Japan, where major end equipment manufacturers try to control the entire supply chain that goes into their products. Another is a TQC system committed to by the entire corporate entity and championed at the top. (We discussed this in Chapter 7, and this is a vital component.) There is also a need for continuous improvement at every level (Kaizen). We first examined this concept back in Chapter 4. In his book *Kaizen*, Massaki Imai said this drive was the single most important source of Japan's corporate competitive advantage. Many of my sources in U.S. subsidiaries of Japanese companies and in Japan agree with Imai.

There should also be cross-functional teams that design products and processes for quality, manufacturability, and *cost*. This is much in vogue today within U.S. high technology companies and is how many

U.S. technology start-ups operate. However, Japanese corporations have developed this philosophy to a fine art. Chapter 5 gives some examples involving Sharp, Matsushita, and Murata Manufacturing. Another element is leadership, or at least parity, in technology with the top competitor. "Technology" is a broad term, covering the spectrum from basic science and innovation to quality and the manufacturing floor. Simply stated, without a distinct and sustainable competitive advantage in one or more of these, the business enterprise is ultimately doomed. Closely linked to technology is superior product, process, and manufacturing engineering. As we've already examined in detail, this is recognized by most observers as where Japanese corporations enjoy an overwhelming competitive advantage over their American counterparts. My own experience in electronics supports this conclusion. A related Japanese advantage is in the use of statistics in design, process control, and problem solving. The Japanese love affair with statistics began with Deming's pioneering work in Japan during the 1950s and has now spread throughout corporate enterprises there. It is also part of the educational system in Japan and represents another competitive advantage for them. Finally, Japanese manufacturing companies are world leaders in the use of JIT (just in time) systems to reduce waste and improve responsiveness. As we will see later in this chapter, JIT is a much broader philosophical concept than just the handling of inventory on the manufacturing floor. It also isn't suitable for every part of a manufacturing system.

On the manufacturing floor, flexible manufacturing systems complement JIT where applicable. If properly implemented, flexible manufacturing systems offer the advantages of being able to manufacture small lots of different but related products on the same equipment and greatly reduced manufacturing cycle times. We will explore these points later in this chapter. World class manufacturing also involves the use of capital equipment to replace labor wherever possible with the primary goal as improvement of quality and reliability. In the United States, this is sometimes viewed mainly as a means of reducing labor costs. In Japan, the primary goal is improvement of quality and reliability. Quality requirements are now so tight in much of industry that in many manufacturing operations even the best trained operators cannot duplicate the ability of well designed machines to operate flawlessly on a continuous basis. The use of capital equipment also demands total productive maintenance (TPM) of equipment. Increasing mechanization and automation requires not only good design, but also machines that continue to work. Proper maintenance of such equipment, or TPM, is a critical function often underestimated within U.S. corporations.

The manufacturing process is also sub-divided into manufacturing cells. As discussed later in this chapter, this has been practiced for many years within the U.S. electronics industry. Newer aspects of this initiative, however, include much greater cross-training within each cell and the creation of customer-supplier relationships between sequential cells in a manufacturing process.

Finally, world class manufacturing involves the enlightened use of human resources, such as cross-functional training throughout the corporation. A "labor force" that understands all aspects of the manufacturing process or, in the broader sense, the entire business enterprise, offers a number of benefits. One is obviously flexibility in being able to move individuals from one job to another. Just as important, if one understands where his or her responsibility fits in the whole, it can be performed more effectively. Such training is also an excellent motivating factor. Another is worker involvement through group activities such as quality control circles (QCC) and an effective suggestion system. Used in conjunction with cross-functional training and intracompany mobility, such worker involvement can bring greatly increased human skills and ideas to bear on the entire business enterprise. However, to succeed participants in such activities must see the fruits of their inputs actually making a positive impact on the corporation. In other words—as with quality—top management must truly believe in and embrace the concepts. Compensation systems should be primarily linked to both corporate and group goals and performance. In the United States, all too often compensation is tied to meeting "individual" goals. However, if not tied to total corporate performance, such an approach can lead to decisions and actions that may not be in the interest of the entire corporation. As an example, imagine a component division that arbitrarily drops a losing product line that is critical to a key customer of one or more other divisions within the corporation.

Just in Time (JIT) Systems

While the JIT concepts were developed in the early 1950s by Taiichi Ohno as part of the Toyota production system, some feel that it can be applied to almost any industry, including electronic components. We will test this hypothesis toward the end of this chapter when we examine the Flip MLC chip manufacturing equipment as a flexible manufacturing system.

Tools that are used in production planning include bills-of-material (BOM) and materials requirements planning (MRP), whereby a sales forecast generates the number of finished units required. This is worked backward through the manufacturing process to determine what is required from each successive station as well as the additional materials input into each. Accurate yield information throughout the process is critical and must be factored into MRP. Depending on the complexity of the manufacturing system, extensive computer software may be required.

But there are several obvious problems with an MRP system, especially in a complex materials processing business such as electronic components. The first relates to variations in yields. If yields in a particular part of the process are too high, excess inventory is created at that station. Worse, if yields are too low, then there is insufficient material to feed the next station and insufficient units are created at the end. To compensate for such a possibility, each station generally builds a little extra to be on the safe side. This is less of a problem if repetitive commodities make up the product line. It is much more serious when high cost specials are being built. Sprague Electric had a continuing problem of this type with its military specialty film capacitor line with hundreds of different part types and average order sizes under 100 units. Another problem occurs if there are rapid changes by customers in their order quantities, either up or down. This is especially true if there is a long production cycle time. Over the years, this has been one of the problems with serving the United States auto industry as the Japanese component suppliers found out early in the game. So there is very real benefit in creating systems that are extremely flexible and have short production cycles.

The JIT system was developed with two primary goals. The first is to reduce waste of all kinds, including excess inventory, excess time or motion spent throughout the system, and defective units. In so doing, cycle times are dramatically reduced, satisfying the second and primary goal which is to create a system that allows efficient and rapid manufacture of small numbers of different products (in the original application, automobiles). JIT works just the opposite of the more conventional "push" system whereby units are automatically transferred from one station to the next, whether required of not. In JIT, no units are moved until "requested" by the next position. The method for doing so is the so-called *kamban*, or label, that accompanies each batch of units or subassemblies as they move through the line. (In the capacitor industry, they are referred to as lot cards.) Supplier performance is also critical to this approach. To eliminate waste, ship-to-stock programs are developed with qualified suppliers whereby they ship directly to the production line without any incoming inspection. And woe to the firm that misses a promise! Another key part of JIT is the use of machines that automatically stop if a problem occurs. At Toyota, this is referred to as *jidohka* or autonomation. One typical example might be an electronic component with a bent lead that jams a piece of insertion equipment. Some years ago I visited a U.S. automotive facility where electronic surface-mount boards for engine-control computers were assembled. It seemed that everywhere I turned a machine was standing idle with a red light flashing on it. There apparently was a major problem with jamming of warped substrates, although I was not convinced that there weren't also difficulties with the machines themselves. Successful JIT—or any manufacturing for that matter—requires machines that work. At the auto plant, the jamming problem was solved, at least short term, by hand assembly and repair in the back room (which, unfortunately, I was never shown). Missing a JIT delivery is bad enough. Shipping faulty parts to stock is the same as committing supplier-suicide. It should be emphasized that MRP and JIT are complementary and not self-exclusive.

Flexible Manufacturing

Manufacture of small lots of different products also requires flexible manufacturing equipment. As with JIT, the first concepts came from the Japanese automotive industry. Deceptively simple in concept, the key is the ability to change tooling or set-ups extremely rapidly. In the case of automobiles, the Japanese can make such changes within minutes rather than the hours required in the average U.S. plant. This allows easy changeover from one model to another and, along with JIT, leads to the ability to efficiently manufacture relatively small numbers of many different variations. The equipment must be designed so that the tooling that differentiates one model from another can be easily slipped in and out. This is a path the United States auto industry could have followed years ago. Unfortunately, mass production techniques were the rule until recently.

Less one be misled, the Japanese tend to mechanize rather than to automate and to take a very conservative approach to equipment design. They feel that incremental improvement is far preferable to quantum leaps. Rather than try and mechanize an entire series of operations all at once, they prefer to optimize each part of the process and only combine operations once the application of each station is completely understood. This approach is shown conceptually in Figure 8-1. In a capacitor manufacturing operation, stations 1 through 4 might be, respectively, preparation of the capacitor section, lead attachment, encapsulation, and test and marking. In the U.S. approach, there would be a strong tendency to combine all four operations in a single mechanization/automation program. By contrast, the Japanese would take the more conservative approach of doing this in two stages. In this manner, the final combination has a higher probability of working properly with minimum downtime.

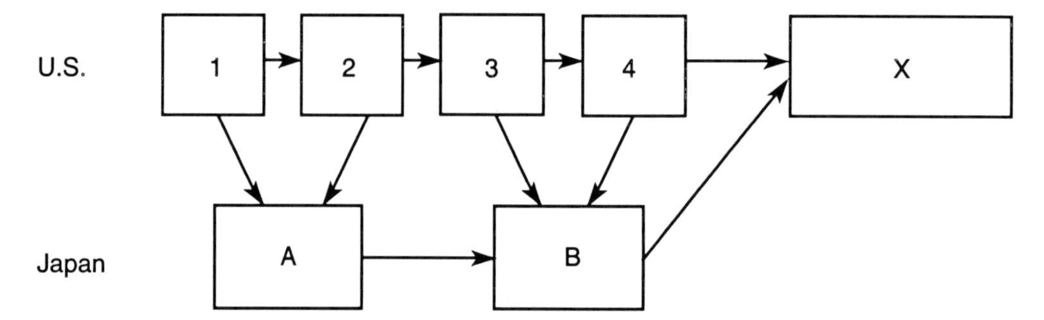

Figure 8-1: Conceptual view of U.S. versus Japanese approaches to mechanization.

I had personal experience with these two philosophies while reviewing the different approaches taken by Sprague Electric and Nichicon-Sprague in the mechanization of a molded, surface-mount solid tantalum capacitor. The assembly and encapsulation part of the process requires attachment of the anode to a lead frame. The Japanese used a conservative design with a limited number of positions on each frame. This simplified both the assembly equipment and the mold chassises. On the other hand, Sprague was more aggressive and used frames with roughly double the density. While making assembly more complex, assuming nearly equivalent yields, the Sprague approach was much more efficient from both a cycle time and capital investment standpoint. No relative information is currently available on yields, the time required to implement the two approaches, equipment downtime, or the ease of maintenance. While N-S has shown interest in the Sprague approach, my guess is that they will continue on their own path, increasing the density of the next production module once they are certain they are ready to. It will take years to truly measure the total return of the two philosophies.

Creating a process that is implemented through use of manufacturing cells has been identified as one of the most important tools in JIT. Movement of parts by cross-trained operators within a cell creates enormous efficiencies by increasing productivity, shortening lead-times, and reducing work-in-process (WIP) inventory. Such an approach makes maximum use of operator skills and involvement and, by creating internal supplier-customer relationships between successive cells, is ideally suited for the JIT system. To be fair, this is how component manufacturers have operated for years. If one thinks about the IC process, the front-end, or wafer processing, is sub-divided into a number of what I would call "cells" such as wafer preparation, mask making (if done internally), epitaxial growth, lithography, wet or dry processing, diffusion or ion implantation, metallization, wafer probe, and then on to assembly, test, and brand. The manufacture of MLC capacitors follows a similar approach with a materials processing front-end, or chip fabrication section, and back-end finishing and test. It is unclear how the chip manufacturing system might be modified to better implement JIT, although it is certainly applicable to the back-end.

It is also not certain that full JIT is for everyone or for every kind of industry. While proponents argue that it is, others disagree, pointing out that JIT often requires additional, redundant equipment. Such investment might be better spent on some inventory. Full implementation also requires total corporate committment, time, expense, complete understanding and characterization of the process, equipment that works, and suppliers you can absolutely count on. (As General Motors painfully learned in 1992, however, a work stoppage or labor dispute anywhere in the supply chain can be serious in a JIT system.) Without all these elements in place, bringing in a team of JIT consultants for several months with the belief that your troubles are over is courting disaster. This is like believing that a new computer is all that is required to fix your accounting problems, improve production planning, and better serve your customers. Still, as the following example demonstrates, even with a complex materials process, intelligent implementation of key JIT principles can have considerable merit.

The Sprague Electric Experience

We have already had a thorough introduction to the Sprague Electric method of manufacturing MLC capacitors back in Chapter 3. In that chapter, we discussed Sprague's curtain—or Flip—build-up machine. This is the same type of equipment that was sent to Mitsubishi, as discussed at the end of Chapter 5. Recalling those discussions, there are two basic approaches to making the capacitor chips. In the first, or tape process (sometimes referred to as "dry"), individual dielectric layers are tape-cast, dried, and stored. Later they are screen printed to deposit the electrode patterns and then laminated under pressure and heat to make the required units. Depending on the capacitor rating, the resultant stack can contain up to thousands of capacitors. This is also true with the "wet" process. The principle advantages with the tape approach are the ability to check each layer for imperfections prior to lamination, and the ability to make small runs. Until the Mitsubishi work, we also felt this process had an inherent yield advantage. The principle disadvantages have been the inability to make and use very thin layers, and high labor content. Continuing progress has been made on the former and mechanization has solved much of the productivity problem.

On the other hand, with the Flip or curtain ("wet") process build-up is continuous until the necessary alternating layers of dielectric and metal have been deposited. While control and yield have been problems, the process is very efficient from a capital investment standpoint and it appears the best suited approach to very thin ceramic layers. This is because individual dielectric films are not handled, since the process involves continuous layering prior to firing.

In many ways the Flip process is an analog of the IC process. That is it involves a repetitive series of steps that, by their very nature, want to make a lot of the same rating of component. This is fine if you are in the commodity business and are selling millions of devices of a few ratings over an extended period of time. However, having finally chosen a niche strategy to try and serve the highly competitive MLC market meant that Sprague needed a much more flexible process and to be able to make smaller lot sizes of capacitors with many different ratings, temperature responses, and performance. After toying with switching to a dry process (we wisely concluded that we were much too far up the learning curve), the answer seemed to be development of a "fast turn-around" Flip. How does one approach such a problem? Really, it requires common sense and a lot of hard work. It is also an excellent example of what is required in a flexible manufacturing system.

Figure 8-2 shows a Flip machine reconfigured from the viewpoint of flexible manufacturing. First , you must have completely characterized and controlled processes. Without this critical first step, there is no sense proceeding further. Second, you must have the ability to rapidly change dielectrics and their associated binders and solvents. In the original equipment, this involved lengthy cleaning of the curtain and pumping systems. In the new approach, these can be interchanged in a matter of minutes with a completely new curtain system for handling the different dielectric paint. While this is being used, the old one is cleaned for subsequent use. The thickness of the dielectric can be controlled at the curtain through the viscosity of the ceramic paint, the rate the carrier belt travels under the curtain, and, most usually, by changing the flow rate of the pump. Continuous sensing of the thickness of the build-up and computer-controlled feedback is a must. If a rush order is received that requires temporary discontinuance of a run in the middle of the build-up, there needs to be a means for storage until the run can be restarted. This is done by shunting the build-up plates into the dryer/storage where they can safely remain for several days. Dryers #1 and #2 tend to be rate limiting steps; this can be overcome by using multiple belts, much like a multiple lane highway.

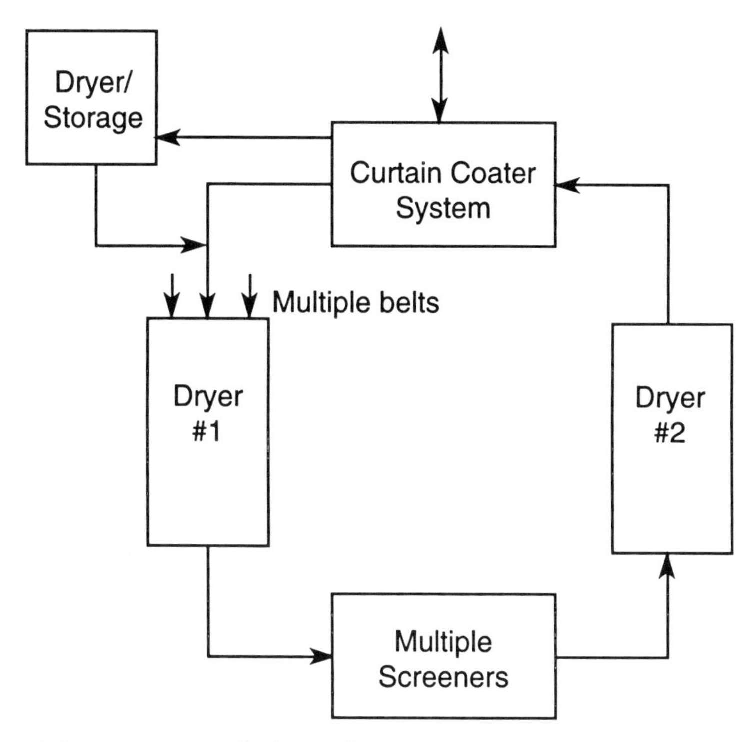

Figure 8-2: "Fast turnaround" Flip machine.

The next set of controls is at the screeners, where the electrode patterns are deposited using conventional thick-film technology. To allow running different ratings simultaneously, multiple screeners are used, either in series or in parallel. The thickness of the metal layer can be varied, either by changing the metal content of the ink or by the mesh of the screen. Further flexibility dictates in-house design and manufacture of the screens. If there is a requirement to go from medium- to high-firing electrodes (for example, from high silver content formulations to platinum-gold), then it is necessary to change both the metallic ink and screener assemblies. This can be accomplished by either using one of the other screeners, or by quick changeover of the screens and squeegee assembly system in the one in use. Through proper design this can be done in well under an hour. Since this a continuous system, all the belts have to be synchronized and their speed under the curtain and in the two driers accurately controlled in relation to the product requirements and the process. The software requirements of such a system are anything but trivial. A nice touch is a visual monitor that shows where all the plates are, much like the control systems that monitor the location of railroad trains.

If the product strategy is purely niche and many small runs of different ratings and performance requirements are required, then such a closed-loop manufacturing system can be made physically quite small. In addition, plate sizes can be reduced so that limited numbers of capacitors are made each run. Since drying must be complete after both dielectric and metal deposition, the rate through the driers in the smaller system must be slowed down accordingly.

What we now have is a closed loop, continuous flow, flexible manufacturing system that solves most of the problems with the original equipment. Since there are redundant systems both at the critical curtain and screener positions, about the only equipment problems that can close the line down exist with the belts, the driers, or, most likely, the computer control system and software. Even JIT deliveries from suppliers are less of a requirement as long as the classical methods of materials qualification continue. Use of high purity, chemically synthesized ceramic powders will probably eventually obsolete these lengthy approaches as well.

Many of the features described above were included in the equipment described back in Chapter 5. Unfortunately, Sprague was never really able to enjoy their benefits since, soon after completing such a machine, they were out of the ceramic capacitor business.

Manufacturing Management

The differences in approach and the general superiority of manufacturing within Japanese corporations, relative to their U.S. competitors in many electronic industries, cannot be due solely to ethnic differences or to luck. Since the success or failure of a corporation is largely determined at the top, we should spend a little time looking at the Japanese chief executive officer, especially as he relates to manufacturing excellence. We have already examined his (and it is invariably a male in Japan) U.S. counterpart in Chapter 6, as well as Japanese managers of their U.S. subsidiaries.

Trying to find an exact model for the Japanese manager is like searching for a needle in a haystack. They are all different individuals, have different educational and work experiences and, therefore, cannot be placed in a single, clone-like mold. Yet there seem to be some striking common characteristics that have shaped them as leaders (that is, in addition to being universally Japanese and male). The Electronic Industries Association of Japan (EIAJ.) publishes a quarterly newsletter in English. In each issue there is an interview with a Japanese leader in electronics. These interviews begin with a brief biography of the interviewee and, using the June, 1989 through December, 1991 issues, I compiled the biographical data that is shown in Table 8-1; if information was missing, I left the entry in Table 8-1 blank. I believe that "commerce" or "commercial science" is similar to a business degree.

TABLE 8-1

Biographical data of selected leaders in Japanese electronics

Name	*Age*	*Company*	*Joined*	*University*	*Discipline*
A. Tanii	64	Matsushita	1956	—	—
T. Sekimoto	66	NEC	1948	Univ. Of Tokyo	Electrical Engineering
K. Kataoka	76	Alps	1948 (founding)	—	—
M. Shiki	68	Mitsubishi	1947	Univ. of Tokyo	—
T. Yamamoto	67	Fujitsu	1949	Univ. of Tokyo	—
H. Sato	63	TDK	1952	Nihon Univ.	Electrical Engineering
H. Tsuji	60	Sharp	1955	Kwansei Gakuin	Commerce
S. Matsumoto	63	Pioneer	1951	Chuo Univ.	Commercial Science
A. Murata	71	Murata	1944 (founding)	—	—
T. Bojo	56	JVC	1959	Univ. of Electro-Communications	Electro-Communications

While it is obvious from this table that—contrary to common belief—not all Japanese electronics CEOs need to have attended the University of Tokyo, there are two important characteristics that these men all share. First is age and seniority. With the possible exception of Takuro Bojo of the Victor Company of Japan (JVC), none can be described as a youngster. (Remember that retirement at the age of 60 does not apply to the executive suite.) Even more striking, however, is the fact that none of these men ever worked for any other company. In each and every case these were their first employers. In the cases of chairman Katsutaro Katoaka of Alps Electric Company and chairman Akira Murata of Murata Manufacturing Company, they were also the company founders. Of these men, the only one I know personally is Akira Murata (as well as his son, Yasutaka, who was recently named president). It is evident that when such men finally reach the top position, they know every aspect of their company. (Contrast this to the United States tendency to hire "best athletes" from the outside, especially when the going gets rough.) But how does all this relate to manufacturing excellence?

All these men entered the business world soon after World War II. They started their careers during a time of extreme scarcity in Japan.

As noted in Chapter 2, the solid-state revolution had started in Western Europe and the United States. With the invention of the transistor at Bell Labs, Japanese industrial leaders in electronics faced a monumental task in trying to catch up. To succeed in business you must have, or be able to develop, one or more competitive advantages. The Japanese are extremely resourceful. Finding themselves far behind the West in basic science and innovation, they chose the route of high volume commodities (such as consumer electronics) and added value through low cost, then quality, and finally performance. Scarce capital for investment became available through a financial system that favored saving rather than consumption. This capital was put to work in the industrial sector in modernization, expansion, and mechanization.

As we have already discussed, higher education in Japan favors engineering over the physical sciences and undergraduate skills in math and science over the masters and Ph.D. degrees. While it is not clear which came first—the chicken or the egg problem—I would guess this engineering orientation was the result, rather than the cause, of the direction taken by industry. Certainly the emphasis on quality that began in the mid-1950s under the influence of Deming and Juran was one of the major reasons that statistics became so important in the undergraduate curriculum.

These Japanese leaders and others like them are fiercely proud men. Since their companies are the core of their life as well as their first and only employers and they have worked in every part of them, failure within the corporation is their own personal failure. Certainly the money is good, as are the perks, the power, and the ability to continue to work long after the regular retirement age. But, if the company goes down, so do they. Contrast that to the United States where "failure" all too often means a big severance check and comfortable golden parachute before landing in another CEO job as someone else's best athlete.

Similarities and a Big Difference

American companies like Sprague Electric and their managers once had several parallels to the Japanese of today. From the beginning, Sprague concentrated on innovation, and manufacturing was an extension of the development process rather than where the innovation skills were concentrated. My father was an engineer, as were many of the original founding team and managers. Like today's Japanese managers, they had a strong personal identification with and stake in the company. Failure carried real financial and personal costs for everyone involved.

In Sprague's early days, they were oriented toward taking processes and products developed in the labs and translating them into high volume commodities. Philosophically—as is still true in many U.S. corporations—the belief was that if feasibility could be demonstrated in the lab then high volume manufacturing was obviously possible. In the early days, high labor content was the rule of the game and performance and reliability were gained through superior skills in materials processing and conservative designs. As a result, Sprague was never terribly expert in mechanization and automation (although this is less true today). Even in the highly successful tantalum operations the capital investment level has been modest relative to Nichicon-Sprague, the Japanese joint venture.

The failure of American managers during the 1950s and 1960s to push for more mechanization and automation seems, in retrospect, to have been a major mistake whose repercussions are still being felt. (Does anyone remember the old Zenith television set ads of the 1960s in which Zenith boasted their sets were wired by hand instead of using printed circuit boards? At that same time, Japanese companies were striving to reduce the "human input" into their television manufacturing.) I had a discussion about this point a number of times with Bob Costello when he was with Delco. Comparing visits to Sprague factories with capacitor operations he had seen in Japan, Bob observed how more highly mechanized they were. He argued that, long term, this would lead them to produce higher quality products because, in high volume manufacturing, well designed equipment will always do the job more uniformly and with higher quality than the most skilled and best trained operators. I was finally forced to agree with him!

So what is the source of manufacturing excellence in Japan? I believe such excellence was *forced* on them by necessity, is consistent with what became the focus of their educational system and ability to add value, and has continued to grow in strength supported by the access that Japanese corporations have traditionally had to inexpensive and ready capital. Fiercely competitive industrial leaders know their companies and, unlike the United States, failure is personal and not the road to another, sometimes higher paying, job. This is in sharp contrast to the recent American tendency to fill the executive suite with outside executives, often with little knowledge of the industry. Skills of such executives, more often than not, are oriented toward the legal and financial. (Ever wonder what an integrated circuit designed by a lawyer or accountant would be like?) While this can often lead to short term improvements in profits—often by cutting back on "waste" like product development and customer support—such backgrounds and talents are not necessarily useful in planning and staffing for the future.

My comments do not mean that there aren't many U.S. corporations that practice management policies oriented toward the long term or whose senior executives know their business and care passionately about its success. (To name just a select few, these include AMP, Intel, and Motorola in electronic components.) However, we don't have enough executives and managers—or boards of directors and shareholders—that take the longer view. We need more.

This chapter is shorter than the two that preceded it, which might seem surprising. However, this is because Japanese success in manufacturing is the summation of the factors discussed in this and the previous two chapters, including the slavish attention to quality throughout a company and the tight bonds between manufacturers and suppliers that make a system like JIT viable. The key points of this chapter and the previous two can be summarized as follows:

- The United States's superiority in scientific and engineering universities is counterbalanced by the superior educational system through high school found in Japan. U.S. employers will have to provide continuing education and strong in-house training for employees.
- Despite continuing improvement in the quality of U.S. goods and services, corporations in this country continue to play a catch-up game. Initiatives such as ISO-9000, Malcolm Baldrige, and Six-Sigma are important contributors to our improvement. Still, all evidence supports the observation that Total Quality Control (TQC) in Japan is still more all-encompassing of the total business ethic than in the United States.
- Japanese corporations tend to form closer linkages with their first tier suppliers than is customary in the United States. With many of the major Japanese firms, this is facilitated by the industrial group or *keiretsu* structure. Penetration of such groups by outsiders, be they foreign or domestic, is extremely difficult.
- The competitive edge that most Japanese corporations have over their Western counterparts in manufacturing appears based on two primary factors: first, the background and one-company orientation of the Japanese CEO has led them to recognize that manufacturing is where the greatest value-added can be generated. Secondly, as the result, this is where they have chosen to concentrate their available financial and human resources.

As a final observation, I believe the U.S. skills in innovation and basic science and those of the Japanese in process engineering and manufacturing are complimentary, not contradictory. We will explore this notion later.

CHAPTER 9

The Marketplace

Most of our discussion so far has concerned itself with the importance of suppliers to the customer. But the reverse is even more important, because with no customers there isn't a business. And there are many customers in the marketplace. As we noted earlier, in 1989 world noncaptive semiconductor revenues were approximately $50,000,000,000, of which 80%, or $40,000,000,000 were in integrated circuits. Capacitor sales in 1989 were roughly $8,000,000,000 on a world basis, with $3,500,000,000 belonging to Japan and the United States market share a stagnant $1,500,000,000.

These numbers aren't insignificant, but they are dwarfed by the total world electronics industry which now approaches $600,000,000,000 (the source of this number will be explored later in this chapter). However, it is necessary to offer several caveats concerning this and other data in this chapter. For one thing, electronics market data on a world basis is notoriously inaccurate; nowhere is it pulled together and summarized in a manner that one can trust with certainty. Secondly, although trade associations such as the Electronic Industries Association (EIA) in the U.S. do the best they can to report accurate information, there are key companies that do not report and whose statistics must be estimated. This is a particular problem with the semiconductor industry. There is always a major difficulty with currency translation. For example, the Electronics Industry of Japan (EIAJ) uses different exchange rates than those that actually occurred in the reported year for some of its international data. Finally, there is uncertainty about whether or not some of the Japanese statistics include offshore value-added manufacturing.

However, while accuracy of data on the international electronics market may be questionable, the overall trends and magnitudes reflected in such data are generally reliable. Therefore, I have chosen to use the statistics as they are reported, with the EIA (*Electronic Market Data Book*, 1988 Edition), EIAJ (*Facts and Figures on the Japanese Electronics*

Industry, 1990 Edition), and Sprague Electric's market research group as the major sources.

Table 9-1 gives world production, exports, and imports of electronic goods in 1988 by major geographic region and was prepared from EIAJ data. "NIEs" means "newly industrialized economies" and is the data for Hong Kong, Singapore, South Korea, and Taiwan.

TABLE 9-1

World production, exports, and imports of electronic goods in 1988
(in billions of dollars)

Region	*Production*	*Exports*	*Inports*
United States	$207	$56	$69
Japan	$166	$71	$10
Western Europe	$153	$95	$124
NIEs	$50	$53	$31

The Western Europe statistics are suspect, since there is no central European agency in a position to accurately report them and currency exchange rates are particularly complex. The NIE statistics also look strange since they show larger exports than production. This gets into the whole problem of accounting for value-added manufacturing, such as the assembly and test of electronic components and end equipment. Nevertheless, the table clearly shows the enormity of the entire market, the negative trade balances of both the U.S. and Europe, the importance to Japan of electronic exports, and how little Japan imports in electronics. Using the same source, Table 9-2 shows companion data, for the U.S. and Japan only, by major market subdivision.

TABLE 9-2

1988 Comparison of Electronics Industry in U.S. and Japan
(in billions of dollars)

	Production	*Exports*	*Inports*
United States:			
Industrial	$150	$38	$33
Consumer	$6	$3	$16
Components	$51	$15	$20
Japan:			
Industrial	$77	$23	$3
Consumer	$33	$17	$1
Components	$55	$30	$7

Table 9-2 documents the huge loss to the United States represented by the Asian dominance of consumer electronics. Our negative trade balance in electronics is primarily in the consumer market and in electronic components, with the principal beneficiary being Japan.

The EIA subdivides the U.S. electronics industry into somewhat different categories as is shown in Table 9-3, which gives the major components of the 1987 balance of trade. The negative total of approximately -$18,000,000,000 is not too far off the 1988 EIAJ estimate of -$13,000,000,000. Only in industrial electronics (primarily computers and related products) was there a positive U.S. trade balance. As we are all aware, this important market segment is also under heavy pressure from foreign competitors, especially Japan. Passive components such as capacitors, resistors, and connectors are included in electronic parts.

TABLE 9-3

1987 U.S. trade balance in electronics
(in billions of dollars)

Communications products	($1.9)
Consumer electronics	($13.6)
Electron tubes	($0.1)
Electronic parts	($4.0)
Industrial products	$3.2
Solid-state products	($1.4)
Total	($17.8)

Since much of our focus has been on electronic components, especially semiconductors and capacitors, we need to show selected statistics here as well. Table 9-4 shows the 1989 world semiconductor market by major geographic region as estimated by In-Stat.

TABLE 9-4

1989 World Semiconductor Market by Major Regions
(in billions of dollars)

United States	$15.0
Japan	$19.8
Europe	$9.1
Rest of the world	$6.0
Total	$49.9

While the figures for Europe and the rest of the world are not small, the United States and Japan clearly enjoy most of the action. In

1984, the equivalent numbers for the U.S. and Japan were $11,600,000,000 and $8,000,000,000 respectively, showing how much the balance of power shifted in the intervening five years. The crossover point occurred in 1986, when non-captive Japanese shipments of semiconductor products first exceeded those of the United States.

Finally, as far as data is concerned, Table 9-5 shows the 1987 capacitor consumption statistics by both dielectric and geographic region. The data was supplied courtesy of Sprague Electric. (PFO stands for "paper, films, and oils.") As is the case with semiconductors, Japan also dominates in capacitors. This is especially true now that AVX is owned by the Japanese giant in electronic ceramics, Kyocera, As a result of this acquisition, Japanese companies not only now control the dominant ceramic capacitor segment on a worldwide basis, they also account for the lions share of aluminum electrolytics.

TABLE 9-5

1987 World Capacitor Demand
(in billions of dollars)

By dielectric:		By region:	
Ceramic	$3.0	United States	$1.5
Aluminum	$2.5	Japan	$2.8
Tantalum	$1.0	Europe	$1.3
PFO	$1.3	Rest of world	$2.2
Total	$7.8	Total	$7.8

So far we have shown the huge size of the world's electronics industry and the increasing dominance of Japan in the consumer segment and in components. In previous chapters, we have shown that Japan's goal is to dominate the world's semiconductor industry and its related materials and equipment infrastructure. This goal relates to a parallel move toward dominance of the computer industry. We have also traced a number of the strategies and tactics Japan is using in innovation and in manufacturing to reach these goals. We now need to discuss what is required to serve both the U.S. and Japanese markets.

Serving the U.S. Market: The Sprague Experience

We have previously seen that Japanese companies work very hard to form close and lasting linkages with their first tier suppliers, and in some cases such suppliers are connected through huge industrial groups known as *keiretsu.* In other cases, top suppliers have reached such a position through excellent products and service, hard work, and sustained effort. (This has also been true of those relatively few United States electronic multinationals that have been successful in Japan, as we shall discuss later.)

This picture might give the impression that somehow Japanese companies are "easier" on Japanese suppliers than on foreign ones. Nothing could be more misleading. Japanese firms are demanding on whoever serves them, be they local or foreign. Service is honorable in Japan and it is unacceptable to serve badly. All too few foreign firms have recognized that getting established in Japan is just plain hard, whether you are an American company trying to break in or a Japanese corporation striving to get established where there are already capable and qualified sources of supply. But, as we shall see, in general the Japanese have worked harder and longer than their U.S. counterparts to become truly global in industries such as automotive and electronics.

In *Competitive Advantage* Michael Porter states that one of the competitive advantages the Japanese have over the U.S. is both a more sophisticated consumer and, at the industrial level, more knowledgeable and demanding purchasing personnel. I'll bet buyers at such top U.S. firms as IBM, Delco, AT&T, Hewlett-Packard., Motorola, and others take serious exception to Porter's conclusion, as should American consumers with their sophisticated taste for Asian televisions, stereo equipment, and automobiles. (Admittedly, with the demise of the U.S. industry, consumers really don't have much choice in consumer electronics.) In addition, it was the American computer industry that turned to Japanese DRAMs when U.S. sources were unable to keep up with demand, only to find them superior in quality and, in some cases, performance. So, is the electronic end equipment industry in the U.S. easier for foreign suppliers to serve than it is in Japan? As we shall see, the answer is yes. On the other hand, it also depends on how individual companies have approached the different markets.

One way to address this question is to look at some of the Sprague Electric experience in the U.S., Europe, and in Asia. In October, 1962, Sprague dedicated its new central research laboratory in North Adams, Massachusetts right across Marshall Street from corporate headquarters. In preparation for this important event, Sprague created a comprehensive brochure entitled *Sprague Electric...an industrial portrait.* Even 30 years later, it is a fascinating document, showing the breadth of different types of sales coverage employed by the then-largest supplier of electronic passive components in the United States and possibly the world. Sprague had chosen the strategy of being a broad-based supplier of not only passive components, but also semiconductor devices, with a major emphasis on reliability and superior performance. Strategy is a key word, since "broadbased" and "leadership" imply very different approaches than the selling of commodities.

With total annual revenues of $80,000,000 in the early 1960s, Sprague still employed a highly centralized organization, with marketing and sales headed by industry veteran Neal W. Welch. Neal had succeeded the early architect of the sales and marketing organization, Julian K. Sprague (brother of founder "R. C.") after Julian became president in 1952. Most U.S. sales were handled by company-employed direct salesman who invariably were graduate electrical engineers. A few very selected manufacturers representatives were used where they had special contacts at major customers. A key strategy was to target market segment leaders in the U.S. and this called for very close working relationships at the purchasing, quality, and engineering levels. Many "specials" designed for companies such as IBM later became industry standards. Specialists in particular component areas served as complements to the line salesmen, a function that became especially important as Sprague entered the semiconductor market. These functions were further complemented by a large group of North Adams-based account representatives who manned the phones, took orders directly, and served as Welch's eyes and ears into the industry. In addition, market research was evolving into one of the strongest in the entire components industry.

Even with the heavy emphasis on direct sales, Sprague was also a pioneer in serving the distribution market through the Sprague Products Company, a wholly-owned subsidiary founded by Harry Kalker in 1933. As a result, back in the early 1960s Sprague Electric employed in the United States most of the sales and distribution techniques used by the components industry today. Sprague was known as an extremely responsive supplier of very high quality, high performance, electronic components, which exacted a price premium and which were seldom received when promised. But did they work when finally delivered! At the time, superior performance and reliability more than made up for the inconvenience of missed shipments.

Unfortunately, this strategy would eventually cost the company dearly as the global competition continued to improve and buyers turned more and more to the purchase of commodity components. Starting in the 1970s, increasing numbers of major customers began to compare and rate suppliers on a total basis, including the meeting of delivery promises. On this last measure, Sprague did not perform very well. We had been slow to develop the necessary internal production control systems, and there was an added problem. When Sprague gave a major customer emphasis—for example, an all-court press to meet the IBM requirements—that invariably meant other customers suffered. Sprague's internal market data in the late 1970s and early 1980s clearly showed that we had the highest market penetration and performed best, in essentially every aspect, for the "Top Three," the name we gave to IBM, Delco, and AT&T. There is no question that these customers received preferred treatment. Not surprisingly, we had a much more difficult time gaining an effective position with the group of major electronic customers categorized as "The Next 35." Strong competitors such as Kemet and AVX did well by concentrating in this area. Our early delivery problems with Sanken were another clear example. We had chosen to honor our commitments in the United States first.

Sprague Electric...an industrial portrait. briefly described Sprague's foreign business in the early 1960s which, in Europe, consisted primarily of export sales and manufacturing to serve U.S. multinationals located there. Again the emphasis was on the major U.S. market leaders. The Asian locations focused on use of low cost labor for manufacturing. At the time, overseas sales were primarily in Europe and there really was no overall strategy for Asia. Japanese component suppliers were not considered serious competitive threats, primarily due to very poor early quality and reliability. Thus, the market potential that eventually existed was not anticipated. Like so many other U.S. corporations, the emphasis was on the huge American market and serving U.S. OEMs as they moved offshore. Even as recently as the latter 1970s, the only serious Sprague efforts in Japan were the Nichicon-Sprague joint venture in solid tantalums and the Sanken semiconductor arrangement that we have already covered.

Corporate Structure and the Marketplace

As Sprague grew in the 1960s and increased its semiconductor efforts, so did the internal pressure to change the organization. It was becoming impossible for a few people at the top to make all the decisions, especially in ICs. One of the biggest problems related to what were known as the product marketing managers—strategic marketers who, among other things, controlled pricing. While the trend was to move them physically to major plant locations, they still reported into Welch's organization. This led to the untenable situation that the plant locations and business units were held responsible for profit and loss while still not having full control of their product pricing. Major conflicts sometimes could be resolved only by the CEO. This was obviously an impossible condition.

After assuming responsibility for the semiconductor division in 1968, I was able to convince management to give us all the tools we felt we needed to operate. In 1969, the semiconductor division became a separate, completely decentralized entity with, among other critical functions, its own independent sales and marketing activity. This was fine while the product lines were commodity logic devices such as the 7400 TTL family. Unfortunately, this strategy proved to be a loss leader. A switch to custom linear ICs for customers such as Zenith, Delco, Polaroid, and RCA created direct conflict at the purchasing level between what appeared to the buyers as rival Sprague sales organizations. All these customers were major users of Sprague capacitors, and the semiconductor division needed the capacitor leverage to get its semiconductor foot (so to speak) in the door. Once in, we were treated as just another supplier of active devices who had to earn its own way. So, swallowing our pride, back we went to a single sales organization under Neal Welch for all types of components, supported in the field by semiconductor specialists.

Regardless of bruised egos—and there were many—the hybrid approach worked. Coupled with the resignations and terminations that resulted, Sprague saved a lot of money at a time when it was really necessary to do so. After years of continuous losses, the semiconductor division turned profitable in the fourth quarter of 1971 and remained that way right into the 1980s, with the exception of the electronic industry disaster of 1975. To someone who has never had profit (and loss!) responsibility, it is impossible to describe the joy we all felt when we finally reached that first full quarter of profitability, especially since it proved to be no fluke and continued for many years.

Since then, the Sprague corporate organization has continued to evolve toward worldwide decentralization on a product line basis. The industry has also moved toward increased use of commodity components with delivery, customer service, and price the driving forces. In capacitors, quality and performance from top suppliers are essentially equal and seldom offer a competitive advantage anymore. With this trend, Sprague has also moved toward less direct sales and more use of manufacturers representatives. The primary aim here has been to reduce cost and to make sales costs variable rather than fixed. How well this serves the customer remains to be seen. These changes also made it organizationally simple to sell the semiconductor group. This divestiture, along with Sprague's recent sale of its equity position in Nichicon-Sprague, removed its last true operational presence in Asia. Today, Sprague's emphasis has returned to the United States and to Europe.

Sprague's history shows how fluid organizations must be to be responsive to changing customer needs and to changes in the competitive environment. *There absolutely is not a single correct organization structure.* But a company's structure should be dictated by the marketplace and customers, not by internal needs or corporate politics. We see too much of this in the United States. One year a company may proudly announce that it has decentralized "in order to be more responsive to our customers" and "to facilitate quick decision making." A year or so later, the same company may report that it has recentralized "to strengthen control in trying times" and "to make restructuring initiatives easier to implement," i.e., to save money. Whatever the change, the financial community generally applauds—at least for the moment. Occasionally such changes are really done to shake up a lethargic business, or at least make the responsible executive appear responsive. This last reason is called the "three envelope syndrome," which can be described as follows: Each manager should have three sealed envelopes containing notes in his drawer, usually given to him by his antecedent. They are to be opened in sequence as the new executive faces successive crises. The first says "blame your predecessor." The second says "reorganize." If there is a third crisis and the wolves are at the door, the final envelope reads "prepare three envelopes."

The United States Electronics Market

Is it easier or more difficult to penetrate the American electronics market compared to Japan's? There are several major differences, all of which favor Japanese companies trying to operate in the United States versus American firms trying to enter the Japanese market.

As a generality, Americans buy value for a reasonable price, regardless of the source. The auto industry is a perfect example. It is no longer low price that draws the consumer; it is perceived value. This is also true in electronic components. Once it was learned that Japanese DRAMs were superior in quality to those from the United States, end equipment manufacturers were perfectly happy to buy them. Eventually, most U.S. sources were gone. There are no American corporate partnerships that come even close to emulating the Japanese keiretsu system, a system that effectively keeps out non-members, unless the product or service offered is unique. In addition, Japanese companies in the United States have little trouble hiring top people over here. As a matter of fact, these days Japanese subsidiaries in the U.S. usually offer greater job security than American firms. Many competent former Sprague executives now work for Japanese firms (although some still wistfully recall "the good old days.") With rare exception, it is nearly impossible for a U.S. company in Japan to hire the best Japanese graduates. Language is also a Japanese advantage in entering the American market. Many have learned English and therefore can be conversant on sales and technical calls. Seldom is the reverse true for American firms entering the Japanese market.

Finally, it is much easier for Japanese companies to buy firms or create subsidiaries in the U.S. than visa-versa. Even with the 1973 relaxation of the Japanese government regulations on foreign equity ownership, purchase of successful Japanese corporations is next to impossible, either because of the very high price/earnings ratios of the Japanese stock market or because they just plain aren't for sale. One major psychological reason behind this is that such firms are looked at as people, and the Japanese don't believe people are for sale.

TABLE 9-6

Advantages Japanese Companies Have in the U.S. Marketplace Compared to U.S. Companies in the Japanese Marketplace

- American consumers and purchasing personnel buy perceived value regardless of the source.
- There is no parallel in the United States to the relatively closed Japanese keiretsu system.
- It is much easier for Japanese firms to hire top U.S. employees than the reverse.
- Japanese businesspeople are more fluent in English that American businesspeople are in Japanese.
- It is much easier for a Japanese company to buy or create a U.S. subsidiary than vice-versa.

The bottom line is that getting established in the United States is certainly easier for a Japanese corporate entity than for an American firm to do the same in Japan. Table 9-6 summarizes the reasons why.

The Japanese are extremely thorough once they decide to make a foreign investment, be it the purchase of assets, creation of a subsidiary from scratch, or setting up a new sales channel. In the May/June 1987 issue of the *Harvard Business Review*, J.K. Johansson and I. Nonaka argue in "Market Research the Japanese Way" that Japanese firms take a much more direct route to market research than used by U.S. firms who, they claim, employ large-scale consumer surveys and "other scientific research tools." In such industries as automotive or consumer electronics, Japanese managers get their information first hand directly from wholesalers and retailers and by detailed analysis of shipments, inventories, and retail sales. Bob Yoshida of Panasonic recently described to me the tremendous detail they developed before deciding to locate operations in the United States. This was generated over several years of market analysis, including detailed contacts with potential customers at the purchasing, quality, engineering, and management levels. The point was to try and understand not only exactly what the customer needed, but to also determine any potential weaknesses with current sources of supply.

This same approach was also employed by Ike Takeda of Nichicon Capacitor who spent a number of years in Chicago as head of Nichicon America. Ike claims that one of the principal weaknesses of many American firms trying to penetrate the Japanese market is their inability or unwillingness to supply what the customers require. A Japanese purchasing executive can get pretty frustrated at hearing, "If its good enough for IBM, why isn't it good enough for you?" (We shall discuss this point more thoroughly in the next section.) Finally, Japanese companies in the U.S. try very, very hard to be involved locally and to act like a U.S. supplier. For example, Smyrna, Georgia-headquartered Murata Erie North America (MENA) is extremely active in both the local community and at the industry level. This includes membership in the Japan America Society of Georgia, the education of Japanese students in the United States, U.S. and Japan student exchange programs, and the EIA. Jack Driscoll of MENA (an old friend of mine from his days at Sprague), has been an EIA governor for many years and just stepped down after several years as chairman of the components group.

All this goes to show that in the United States Japanese firms try to be one of us. This is a strategy used by all too few U.S. companies in Japan—except the successful ones.

None of these points should sound new or novel to superior American electronic component suppliers, at least as far as serving the U.S. market is concerned. These are all issues successful vendors have had to deal with in this age of severe competition and shrinking sources of supply. The customer has always been the best source of information on what is required. Increasing numbers of major OEMs have what are known as preferred vendor programs, a parallel to the Japanese top tier supplier relationships discussed earlier in this book, with single or dual sources on key commodities. They have found it much more efficient to work with a few sources than to try and manage many, especially with the move to JIT manufacturing systems. In other words, whether you are a U.S., Japanese, or whatever supplier, you have to be either "the best in class" or very close to it to even stay in the game. That means superiority in quality and reliability, product performance, customer service, competitive pricing, and—tougher for many U.S. firms—on-time deliveries. In the past, this has been true for years with companies such as IBM and a limited number of others. Today, this is required with almost all major electronic end equipment manufacturers. (If the requirement for such overall superiority sounds like a "motherhood" or "apple pie" statement, then you still don't understand the concepts of TQC or Six-Sigma discussed previously.)

TABLE 9-7

Key Requirements for Serving the U.S.—Or Any—Market

- Understand and supply what customers want and need.
- Be "best in class" in quality and reliability, product performance, customer service, and delivery performance at competitive prices.
- With major customers, have direct contact at all key levels, including purchasing, quality, engineering, and management.
- Be perceived as "local."

While Japanese companies generally have an easier time getting a shot at the U.S. market than American companies do in Japan, once Japanese companies are given the chance they work very hard to do the job right. Table 9-7 summarizes the key elements to serving the U.S. (or any) market.

Having just said what seems obvious, we need to consider one more very important point that indicates that things aren't going right for American companies even in the U.S. We have previously cited a *Technology Review* article titled "How Japanese Industry is Rebuilding the Rust Belt." In it the authors, Martin Kenney and Richard Florida, point out the massive investment Japanese auto manufacturers have made in the United States to support their auto assembly plants here and across the border in Canada. These include investments in steel works, rubber and tire plants, assembly factories, and close to 300 auto parts suppliers. Total investment, including the parts suppliers, is more than $30,000,000,000. As a result, some 130,000 jobs have been saved or created, making up for at least some of those eliminated by the "Big Three" U.S. auto companies and by their suppliers. One out of every five cars manufactured in the U.S. now comes from these Japanese subsidiaries, with reported quality at least equal to Japan. While this sounds great for the U.S. economy, there is a flip side. According to Kenney and Florida, the Japanese auto companies have been forced to create their own parts supplier infrastructure because they claim they found few American firms that could respond effectively to their stringent requirements in quality and delivery.

But this conclusion is hotly disputed by several U.S. firms. Bob Marlowe, vice president of marketing for the Sprague Solid Tantalum Division, flatly disagrees with the inferior quality and customer service argument. Despite an excellent position with nearly all U.S. OEMs, the tantalum division has found it almost impossible to penetrate the U.S. subsidiaries of such Japanese companies as Oki, Mitsubishi, Matsushita, Toshiba, and Fujitsu. Parts suppliers are specified in Japan and, in general, end up being Japanese suppliers. Qualification tests are also run in Japan. It isn't that the U.S. capacitors fail; they just never seem to get qualified. Even when an alternative supplier is Nichicon-Sprague, and the products are identical, N-S gets the business. This is a direct example of what is happening to the U.S. electronic components infrastructure as Japanese end equipment companies continue to expand manufacturing in the U.S.

It appears that the Japanese are following the same philosophy they have done previously relative to the semiconductor industry infrastructure in materials and manufacturing equipment. That is, they wish to maintain maximum control over the entire value-added chain, clustering key support industries wherever possible. This is also consistent with the keiretsu system, which they have been exporting to the U.S. as well. Whatever the reason, if this trend continues, there will be more Japanese buyouts of such support companies and lost business to those who aren't purchased. It seems especially ironic that the very reasons U.S. companies are having problems penetrating Japan are beginning to haunt them at home as well. This also makes a strong argument for creating tough local content requirements on such foreign subsidiaries.

Serving the Japanese Market

As was true in the U.S., there is also a lot that can be learned relative to this subject through analysis of Sprague Electric's successes and failures in Japan. In the early 1970s, Sprague began its first operational thrust into Japan with the creation of a joint venture to manufacture solid tantalum capacitors, Nichicon-Sprague (N-S). The Japanese partner was Nichicon Corporation, a worldwide supplier of miniature aluminum electrolytic capacitors. All the technology for the joint venture was supplied by Sprague while financing, creation of the business, and management were supplied by Nichicon. All product sales were through the parent companies. Nichicon controlled all sales in Japan, and, on a non-exclusive basis, in Asia. Sprague controlled all sales in Europe and in the United States. Equity was 50% each, and this was less of a problem than one might expect.

The first ten years were rocky, with heavy start-up costs and losses, up-and-down changes in equity, and a particularly unpleasant failure to expand the product base into other passive component areas. Nevertheless, by the time I joined the N-S board in the late 1970s things were beginning to perk and the joint venture had become an important manufacturer of dipped tantalums for its owners. Later, this would expand into surface-mount devices as well. When the joint venture was first formed, Sprague was the leading capacitor manufacturer in the world. The future potential of the Asian market was relatively uncertain, except in consumer electronics. Since Sprague management still considered Japanese competitors as a minor threat, there seemed little to be lost in such a venture. Sprague controlled the then two largest markets, the United States and Europe, was required to contribute no financing, and made its contribution through technology. Not a bad deal! In the context of the time and competitive situation, it is a credit to the tantalum division management that they persevered in even creating the venture and that corporate management supported them.

Nichicon's goals seemed quite clear: acquire the technology, gain access to the European market through the Sprague sales organization, expand their market penetration in the United States (again through Sprague), and establish a long term relationship with the world leader in capacitors. Gaining an initial foothold in the American market was not really an early issue, since Nichicon was already firmly established with a number of important customers in aluminums.

Looking back, our initial goals at Sprague seemed to be learn about the Asian market and to gain a long-term position in Japan which we felt might pay off someday. The latter is one thing we really did not accomplish, at least as far as the Japanese customers were concerned; to them, the supplier was still Nichicon and not Sprague. Still the joint venture had a very important advantage as opposed to a straight technology license: Sprague prevented N-S from serving the markets outside of Japan and the rest of Asia except through the Sprague sales organization. In other words, the customers continued to deal with Sprague. In the case of straight licensing, it would not have been very long before Nichicon became a direct competitor to Sprague in tantalums in all the world markets. That is what is now starting to happen since Sprague's sale of its N-S equity to Nichicon in July of 1991. Using their established position in miniature aluminums, Nichicon is now beginning to bring the N-S tantalum product directly into the U.S. and Europe in competition with Sprague, something the original joint venture agreement prevented.

On the other hand, the N-S venture did not give Sprague direct access to the Japanese customer base. This continued to be served by

Nichicon. In protecting our main markets, we also failed to directly penetrate theirs. Unlike the situation in the United States, where Sprague only controlled tantalum sales, Nichicon was chartered to sell other Sprague passive components in Japan (although the related revenues were relatively minor). Nichicon claimed this was because Sprague's primarily industrial oriented capacitor families were incompatible with the Japanese consumer market. This argument made little sense in recent years as the Japanese industrial market surged. Strategically, Sprague violated one of the first principles of doing business in Japan: be perceived as local. In hindsight, this was unfortunate since Sprague had a superb reputation in Japan as one of the world's leading component suppliers and, with direct customer interface, could have made a major penetration into the marketplace in other passive component families even if tantalums continued to be sold by Nichicon.

Still, over the years N-S offered Sprague a number of benefits. It gave us access to market information in Japan not easily available from other sources as well as access to Japanese technology, especially in manufacturing engineering and equipment. In addition, Sprague received product to sell from Nichicon with no capital investment required, and royalties and dividends were in the six figure range by the mid-1980s. There were also modest equity profits when the company turned profitable. When Sprague sold its equity position in 1991, it received $18,000,000 in cash. On their side, Nichicon finally gained full control of a world class, profitable solid tantalum capacitor business.

All this brings us full circle to where this book began: the Sprague semiconductor relationship with Sanken. Here Sprague wanted a sales channel for semiconductors in Japan, having failed in an effort to go direct. Unlike capacitors, Sprague had only a minor reputation in semiconductors and it seemed that an established Japanese firm such as Sanken was ideal. On their side, Sanken wanted the state-of-the-art Sprague power IC line and, eventually, the technology as well. As with Nichicon, the Sanken relationship gave the Sprague Semiconductor Group no direct access to the Japanese customers. When Sprague and Sanken parted ways in the early 1980s, Sprague was back to ground zero as far as serving the local market in Japan was concerned.

The delivery and visual quality problems we encountered with Sanken were all further clues to what is required to be successful in Japan. When we first introduced the interface driver line in the U.S., we were seldom penalized for late deliveries because the product was so new and offered so much "value-added." Not so in Japan. To these customers a missed delivery was a missed commitment and unacceptable—period. Sprague also learned that acceptable quality in the United States may still be unacceptable in Japan. For example, visual defects indicate two deficiencies to a Japanese customer: poor workmanship and the inability to meet an agreed-upon specification or committment. Today, such defects are also becoming unacceptable to most major U.S. OEMs.

Those mechanical and electrical failures that did occur required immediate failure analysis, a requirement we were never able to satisfy in a timely fashion out of the United States. This was solved only when Sanken created this capability in-house in Japan and started to "Japanize" our product for the local market. Similar problems with all too many U.S. semiconductor suppliers has led most Japanese customers to distrust American sources of supply and to develop a strong preference for local sources. This has made penetration of that market extremely difficult. By analogy, imagine trying to force an IBM or AT&T to purchase parts from what they feel are unreliable suppliers.

Still, Sanken would never have purchased the Sprague Semiconductor Group in 1990 if they hadn't concluded it offered a strong overall capability and had developed a commitment and ability to meet stringent customer service and quality targets—or could be "trained" to do so!

There is an interesting indirect sequel to the Sprague-Sanken break-up that occurred in the mid-1980s and which provides further insight. I was frustrated with the results from our direct sales office in Tokyo—SJKK—and recognized how really tough it was for an unestablished U.S. firm to hire outstanding talent in Japan. As a result, on one of my frequent visits to Japan I decided to try and hire a Japanese heavyweight to head-up our overall semiconductor efforts there. My target was Dr. Denda, my old associate from Sanken, who we have already met in the first chapter of this book. I do not recall whether he was still at Sanken at the time I decided to hire someone like him or whether he had moved on by then. We met in the lounge of the Tokyo International Hilton, one of the great locations in Japan for people-watching. I especially like the "who can bow lower" charade that seems to go on endlessly between senior business executives and their subordinates. Repeatedly, I have seen older executives bow lower and lower so that the younger man—who can never be higher than his superior—ends up with his face almost on the floor.

Denda-san is a slender, good looking man with an excellent grasp of the English language and a warm smile. He is also particularly perceptive and, like most Japanese executives I have met, extremely bright. After pleasantries about my trip, business in general, and a friendly update of the Sanken situation, I rather nervously began to explain the reason for my visit. Actually, he was well prepared since one of our local contacts had already met with him. He listened politely and then began to explain just what he felt was required for Sprague to be successful in Japan in semiconductors and how long it would take. In addition to direct sales engineers and support personnel, we would need all the value-added services that Sanken had developed plus a local design and applications capability. I do not remember the exact staffing requirements, but they were staggering. Nor were the equipment needs trivial, although many could be satisfied by excess gear that already existed in the States. It would take a minimum of at least three to five years to breakeven and, in the interim, annual losses would be in the six to seven digit range. Making a mental calculation of the impact this would have on the overall Sprague results and how Penn Central, Sprague's owner at that time, would view such a project, I knew the issue was dead. After further discussion we agreed to talk again and parted somewhat more coolly than we had met. Due to the limited commitment we were willing to make, we never were able to hire a top person. Although we corresponded periodically, Denda-san and I did not see each other again until my return to Tokyo in June of 1991. As always, he was extremely helpful even after all the time that had passed.

American Successes in Japan

Are there successful U.S. companies in Japan and, if so, how have they done it? The answer to the first part of the question is "yes" and the answer to the second part is "they have done it the old fashioned way: they have earned it!" Some of the most successful have built on early established positions developed many years ago, while others have penetrated the market more recently through consistent and continuing effort. All are world leaders in one or more important product areas. For example, AMP is far and away the world leader in connectors with total 1990 revenues exceeding $3,000,000,000, close to five times those of their next largest competitor, Molex. AMP holds the largest market share in every geographical region, including Japan where they are considered to be as much a Japanese company as any of their local competitors. They hire the best people, design products specifically for the Japanese market, and are active in all the industry trade associations.

AMP started to position itself in Japan shortly after World War II as a "yen company," one of the only investment routes available at that time. Only when the yen became convertible to foreign currency in 1964 could the earnings generated through this initial investment be repatriated. In the meantime, they had all been reinvested in Japan. By 1964, AMP was well established with a 100%-owned subsidiary. They have continued to build on this position ever since. Very few other United States companies took advantage of this early opportunity because of the high risks involved.

When the yen became convertible in 1964, penetration became much more difficult as the Japanese government introduced extremely tight regulations relative to foreign investment in Japan. There was an absolute prohibition on non-Japanese equity ownership exceeding 50%. (Contrast this stance on foreign ownership with that of the United States government!) However, Texas Instruments was able to overcome this restriction by using its basic patent position in ICs as a bargaining chip. The most important was the Kilby patent mentioned in Chapter 2. Unwilling to provide licensing to Japanese semiconductor firms without receiving permission to make a 100% ownership investment, a compromise was finally reached after lengthy and acrimonious negotiations. Through an interim joint venture with Sony, Texas Instruments was able to make its investment and the Japanese to get the necessary licenses. However, getting going was very difficult, partly because of the scars left by the negotiations and partly because Texas Instruments was definitely viewed as a non-Japanese corporation. Despite this, Texas Instruments continued to invest and target Japan as a major business opportunity. Like AMP, they were able to hire good people, starting at the top. Today Texas Instruments has a major position in Japan with research and development laboratories and major operating plants producing the most advanced ICs. Quoted in the April 6, 1991 issue of *Electronic News*, Texas Instruments chairman Jerry Junkins reported that their annual sales in Japan were about $1,000,000,000. They were also building a new 8" wafer fab in Hiji, Japan, and Junkins was quoted as saying he "wanted more."

Why was Texas Instruments able and willing to take such a stand when other U.S. firms with equally basic patent positions, such as Fairchild, were not? Fairchild and others were willing to license their technology without a quid quo pro other than the immediate financial return that came with license fees. But Texas Instruments held out for direct investment. One answer is an enlightened management. Certainly Texas Instruments recognized the market potential and was willing to take the risk. However, since I know some of the executives involved, I also believe that the war was still fresh in their minds and, with typical Texas toughness, there was no way they were going to accept only license revenues for their patents from the vanquished Japanese!

If other U.S. IC firms such as Fairchild had taken the Texas Instruments position relative to licensing their technology in Japan, might the world competitive situation today in ICs be more favorable to the

United States? Frankly, I doubt it. While the Japanese rise to dominance in semiconductors might have been slowed some, I still feel it probably would have only delayed the time a little before they reached the status they enjoy today. In addition, even with its early start in Japan and the basic market position Texas Instruments has been able to build there, they have been having overall financial problems in recent years. In other words, success in Japan doesn't necessarily mean world leadership.

One of the great success stories in Japan, and in the world, has been IBM. Nippon IBM, a wholly-owned subsidiary, is one of the most profitable business entities in Japan and grew out of an early prewar position held there by IBM. However, postwar introduction of IBM computers and operations into Japan didn't come easily and was possible only after, like Texas Instruments, IBM was willing to license its technology to local companies. IBM has continued to invest in Japan as it has around the world.

A number of other U.S. firms, with equally strong prewar positions in Japan, let what could have become a major opportunity slip away. Prominent names include the U.S. automotive giants, General Motors and Ford. How different that industry might be today if they had persevered as did AMP, Texas Instruments, and IBM. (It is interesting to note that one of MITI's reported "failures" after the war was its inability to create consolidation within the Japanese auto industry. MITI felt this was necessary due to the overwhelming strength of the U.S. industry at that time.)

The above success stories all represent enlightened capitalization on relatively early strong positions in Japan. But there are more recent successes as well. When I visited EIAJ in Tokyo last year, we discussed which U.S. electronic companies they felt were most successful today and why. In addition to AMP, Texas Instruments, and IBM, they also mentioned Motorola and Intel. The reasons given for their success were all similar: leaders in their product areas, strong direct contact with their Japanese customers, forceful local participation in trade and other business associations, excellent people (both Japanese and expatriates), and continuing investment and commitment. In other words, in addition to being product and market leaders, they also acted as if they were Japanese firms. If AMP, Texas Instruments, and IBM all had early presence, Motorola and Intel are both relative newcomers, at least in those areas where they now are leaders.

Motorola is a particularly interesting case because today they lead in mobile communications, especially cellular telephones, and in certain semiconductor segments such as microprocessors. On the other hand, Motorola's early leadership position was in consumer electronics where they had a presence in Japan as far back as the 1950s when they helped rebuild the local economy. Some years later, faced with strong Asian competition, they sold their Quasar television business to Matsushita and exited the consumer electronics business. So, despite an early Japanese experience, Motorola's current position is the result of an entirely different product thrust.

Intel represents one of the true success stories in the American semiconductor business. Founded in 1968 as a Fairchild spin-off (see Chapter 2), Intel today is best known for invention of the microprocessor and a continuing leadership position in this critical segment. If Japan dominates the DRAM business, the United States—led by Intel and Motorola—continues to lead in the design-driven microprocessor field. Intel has apparently done almost everything else right in Japan, although product leadership is certainly their key lever. It will be interesting to see how well they maintain this as Japanese semiconductor firms accelerate their resource commitment to this lucrative market segment.

Lest one conclude that the real keys to serving the Japanese market are product leadership, local presence, and hard work, it must be pointed out that there are also some real barriers, most of which have already been mentioned before, which are mirror images of the factors that make penetration of the U.S. market easier (see Table 9-6). However, none are due to restrictions by the Japanese government. In general, Japanese corporations like to buy from local suppliers and have greater confidence in their ability to meet the stringent quality and delivery requirements than foreign suppliers. This relates to the whole question of "clustering" introduced by Porter in *Competitive Advantage.* This means that if you have a true leadership product, as Sprague did in its Sanken program, Japanese firms will buy from you—at least until local competition can become effective. Above all this means that, in addition to leadership, you must have, or gain, a strong presence *in* Japan.

Penetrating the Japanese Market

As already pointed out in our discussion on relations with suppliers, strong vertical linkages between suppliers and customers are even greater barriers when trying to supply to customers within a keiretsu. In the case of end equipment manufacturers, each such industrial group usually includes a full complement of what are—to a degree—captive parts suppliers. It is hard even for a Japanese supplier who is not a member to surmount such a barrier. This presents a real problem with the Japanese computer industry, where each major manufacturer is a member of a different industrial group. Trying to force NEC, Toshiba, or other computer firms to buy components outside their group—unless not available within the keiretsu—is like telling their equivalents in the United States who their suppliers must be. Good luck! This is one of the main problems with the recent U.S.–Japan semiconductor trade agreement and its target 20% U.S. share of the Japanese market. While keeping the pressure on, how do you influence these massive business conglomerates? It is often said that nations don't compete in the business arena, only individual companies. But with a keiretsu one is fighting something close to an economic empire.

You also have to have the products available. In a different view concerning the above trade agreement, my friend Dr. Denda told me that, because of the agreement, there was enormous pressure on Japanese managers to "buy American." However, he said that Konica and manufacturers of other consumer electronics gear had found it almost impossible to locate the capability to design and manufacture the unique custom circuits required outside of Japan. To serve this niche requires a major commitment over a lengthy period, all locally. Texas Instruments chairman Junkins recently told a Senate subcommittee that while 20% is "essential," it will not be obtainable without significant purchases of custom semiconductor products. Who is going to supply them? If any U.S. company has the courage and willingness to try, I bet that companies such as Konica would be willing to give them the opportunity.

Earlier we discussed the investment Japanese auto companies have been making in the U.S. In an August 6, 1991 Boston *Globe* editorial in their "Opinion" section, Lester C. Thurow, dean of the Sloan School at the Massachusetts Institute of Technology, pointed out that while Japan can freely buy or build businesses around the world, the reverse is not true. As example, in 1989 Japan spent $254,000,000,000 in foreign investments while there was only an equivalent of $16,000,000,000 direct foreign investment in Japan. The reasons given for this imbalance were Japanese government restrictions (which officially ended in 1973), the very high price/earnings ratios of Japanese companies, the keiretsu system, and the fact that good Japanese companies are just not for sale. If this continues unabated, Thurow argues, that the entire world free market system will come unglued. As far as this discussion is concerned, this only further points out how hard it is to establish a presence in Japan through direct investment, leaving joint ventures, "partnerships," or abandonment as about the only, admittedly much less attractive, alternatives. (That is, unless you are already there or an absolutely unique company such as an Intel or a Motorola.)

TABLE 9-8

Key Requirements for Serving the Japanese Market

- Total committment by the CEO, senior management, and the board of directors for a sustained allocation of resources for a minimum of five years (it can easily take longer).
- Sustainable, worldwide product leadership that is designed for, and effectively serves, the needs of the local market.
- An understanding of local customer requirements and a willingness and ability to meet them, even when some seem superfluous.
- A total quality system targeted at zero defects.
- Delivery performance that is compatible with customer just-in-time requirements.
- Competitive pricing.
- Strong local presence, including direct contact with key customers at all levels, including purchasing, quality, engineering, and management.
- The best local people available. A Japanese heavyweight at the top is a must.
- The patience of Job, or at least of your best local competitor.
- If all else fails, find a partner or punt.

So whether or not you are already established in Japan, if you are committed to be a factor in that market, what elements are required to give you the greatest chance of success? Many are those already shown in Table 9-7, which gives some of the key requirements for successfully serving the U.S., or any market. These are restated in Table 9-8, and selectively augmented to better present what are the unique requisites in Japan. One thing is clear: the job is much more difficult in Japan than it is in the United States.

I received a telephone call several months ago from a former associate who had recently created a new, specialized IC company that was concentrating on custom linear IC for the telecommunications market. He was eager to create an early position in Japan and wondered how he might best get started, especially since, he claimed, he had no real direct contacts in Asia. During several meetings over the next two weeks I asked him to consider an important question before he went any further. First, and most important, he had to decide just why he wanted to try and crack Japan so early in the growth cycle of his company. What was he looking for, licensing income, an expansion of his embryonic customer base, a partner, or what? I argued that it was a big step, especially since there was so much he and his team had to first accomplish in the U.S. He stubbornly countered that he wanted to get into the market while his design capabilities were new and unique and before there was any real Asian competition. It all sounded very much like the reasons Sprague had for making its early semiconductor agreement with Sanken. Even when I explained all the difficulties involved and the time and effort required, he stuck to his guns.

Since he had obviously made up his mind, I asked if he and his team were actively involved in writing papers for publication and presentation at technical conferences such as the Electronic Components and Technology Conference (ECTC), assuming this could be done without revealing important trade secrets. Most such conferences have a strong international audience and, if the work is important and unique, I said he would quickly find himself the pursued rather than the pursuer. He agreed this might be a good way to get going, assuming the very limited staff could find the time. Writing a meaningful technical paper is no trivial task. However, it is a very good way to get recognition both overseas and in the United States.

Next, I challenged his statement that he had no Japanese contacts. Didn't he have friends and associates who now worked for Japanese subsidiaries in the United States? Weren't there Japanese-funded professorial chairs at his university? Didn't he have other friends in industry who might be willing to help make initial contacts? Had he covered the technical journals in the field and books dealing with the electronics industry in Japan to seek out possible names? He agreed that these were all possible leads that he would pursue when he had the time. I reiterated that time, and lots of it, is what the whole process would take.

I asked if he had ever had any dealings with trade associations such as EIA or SIA? To this he replied that he really didn't see much use in such organizations. However, when I pointed out how helpful the EIA had been to me over the years, that it sponsored periodic industrial visits to Japan, and that there were a number of Japanese U.S. subsidiaries who were members and who would be happy to help setup contacts, he agreed to look into membership. Trade associations, and especially the EIA with its diverse membership, can be very effective means of networking. A $1000 associate membership can be a very good and inexpensive investment. I also suggested he might try the American Management Association (AMA) headquartered in New York and see if he could make contact with anyone on their international council.

As another avenue, I mentioned that Tokyo-based Techno-Venture Co. Ltd., was one of the very rare Japanese venture capital firms and that its U.S. subsidiary, Techno-Venture USA, was headquartered in Boston. If he was seeking both equity and a Japanese partner, this would be an excellent place to start. I also mentioned that, although they were probably too expensive for a start-up, the major accounting/consulting and consulting firms such as Peat Marwick, Arthur Andersen, Ernst and Young, McKinsey, and BCG, to name just a few examples, all had Asian offices which should be happy to set-up contacts in Japan (for a fee of course). Probably more suitable would be IVEX International, a consulting firm headquartered in Nashua, New Hampshire, that specializes in creating distribution channels and alliances in Japan.

There is no question that, perhaps more than anywhere else in the world, personal contacts are one of the principal ways things get done in Japan. Last I knew, my friend was busy tapping his own network—which proved much larger than he first imagined—and was in early discussions with Gene Lussier of the Components Group at EIA about joining as an associate member.

We have covered a lot of territory in this chapter, but the most important conclusion is that it is much easier for Japanese corporations to access the U.S. market than the reverse is true in Japan. There are a number of reasons for this. The U.S. consumer and industrial purchasing executive tends to buy perceived value, regardless of the source. But because of recurring problems with U.S. sources of supply—including inferior performance, quality, and design for the local market—Japanese consumers and buyers have acquired a strong preference for Japanese sources. This is true both in Japan and with Japanese subsidiaries in the U.S. This preference is further strengthened where industrial groups or keiretsu are involved.

Japanese have superior language skills and are better able to hire superior personnel in the United States than U.S. companies can in Japan. Japanese corporations work harder at being perceived as local in the U.S. than U.S. companies do in Japan. This is facilitated by the ease of direct foreign investment in the U.S. compared with Japan.

In short, for a variety of reasons it is easier for a Japanese company to enter and serve the American marketplace than it is for a U.S. company to do so in Japan. What should a U.S. company that wants to sell in Japan do? Unless you have the money and product advantages of a Motorola or Intel, I would strongly suggest that a joint venture with a Japanese company be pursued.

CHAPTER 10

Government and Capital

What the U.S. government's role should be—or could be—relative to businesses competing against Japanese firms has long been a subject of heated debate across the political and economic spectrum. The subject of the availability and cost of capital to U.S. firms hasn't attracted the same amount of attention, but it is almost equally controversial. Since the cost and availability of capital is often intertwined with government policy (such as the Federal Reserve's current discount rate), we will examine both topics in this chapter.

The Costs of Government

Both the American and Japanese governments impose costs upon businesses in those countries. The most obvious cost is income taxes on corporate profits. But what about the costs of government-mandated programs? On the question of the comparative costs to business of social or benefit programs, it is difficult to make a direct comparison between the U.S. and Japan except in certain areas. Jim Matsui of Allegro Microsystems recently provided me a Japanese survey showing average benefit costs as a percentage of payroll in Japanese industry. The total was a startlingly low 14.4% for all industry and 15% for manufacturing. The equivalent for a mature U.S. electronics company with a large number of long-service employees is 35% to 40%. However, two major segments, "leisure time" and pension payments, are not identified separately in the Japanese statistics and are included in the payroll numbers for each company. I also was unable to extract these numbers from the Towers Perrin report. However, noting these difficulties Table 10-1 provides a summary of the Japanese statistics for all industry and compares them with a "typical" mature U.S. electronics company.

TABLE 10-1

Comparison of Social Costs in the United States and Japan as a Percentage of Payroll

Category	*U.S. %*	*Japanese %*	*Statutory in Japan?*
Leisure time	9	in wages	Mainly non-statutory
Medical/dental	10	3.8	Mostly statutory
Pensions	3	in wages	Non-statutory
Payroll taxes (including social security)	10	5.2	Statutory
Workman's compensation	5	0.9	Statutory
Housing allowance	—	2.5%	Non-statutory
Other	1	2	
TOTALS	38	14.4	

While exact statistics on leisure and pensions are lacking, I believe that both are considerably lower in Japan, especially in the pension area. For example, Towers Perrin examines retirement income as a percentage of final cash compensation for various job categories in the United States and Japan. For a factory worker in Japan, this figure is 65% as compared to 96% for a U.S. factory worker. For a Japanese accountant, the figure is 53% and 77% for an American accountant. Much lower medical and dental costs, as well as lower costs for "social security" type of programs, give Japanese corporations a real competitive advantage versus the United States. However, housing allowances are an important element of compensation in Japan and essentially non-existent in the United States.

The U.S. government also funds many more social programs than the Japanese government does. Some of these are a result of the social and cultural diversity that is an American advantage, while others (such as those for drug and urban rehabilitation) are the result of our national failures in some areas. All of these programs require taxes and, increasingly, government borrowing to fund. Keeping corporate taxes low and minimizing social costs to business certainly reduces their overall costs, should make businesses more competitive and, in so doing, create more jobs. In addition, reduction of the individual tax rate should also favor increased consumption, which would help businesses. However, failing to address serious social problems through government programs and spending could exacerbate those problems and adversely affect our overall competitiveness. Such social problems are generally interrelated and cannot be solved piecemeal. For example, the solution to education problems also involves solving problems such as drug abuse, childhood nutrition, economic opportunities for students' families, etc. Solving these "macro level" problems can only be done with federal help which, by definition, requires tax income. One of the ironies in this country is that, while virtually everyone feels taxes are too "high" no one seems to know what level is "right." Higher taxes to meet social needs or lower taxes to stimulate business and the economy? There are tradeoffs required for either option and no easy answers. As we will see later in this book, Japan has its social problems today and looming ones tomorrow due to the aging of its population. Yet their problems do not begin to compare to those of the United States.

Government Industrial Policy and Direction

In our frustration as a nation over trade problems, it is easy to look for scapegoats. "Japan, Inc.," that amorphous entity cited by Houghton and others is a popular one despite, as we have seen in previous chapters, the lack of evidence to support the concept. However, the notion of an evil conspiracy where the Japanese government and cartels of businesses work in concert to wipe out entire sectors of foreign economies is a tempting way to discount our own deficiencies. But it is also mostly rubbish. Has the Japanese government practiced protectionism and targeting? Yes, but so has the United States. What about the bail-outs of Lockheed, Chrysler, or, more recently, failed financial institutions? How about the massive procurement of hardware and services by the Pentagon that has created the military-industrial complex in this country? Research and development expenditures in the United States exceed the combined totals of Japan, Germany, France, and the U.K., and 50% of these are funded by the federal government, with close to two-thirds of this aimed at defense. If that isn't a subsidy, I don't know what is! As far as the "Japanese conspiracy" is concerned, while the government certainly aids industry in a number of ways, the fierce competitiveness of the individual companies is such that they care more about wiping out each other than an overseas competitor.

But if all this is true, where did the "Japan Inc." myth come from?

Probably its origin is really the attack on Pearl Harbor, even if that occured fifty years ago and, as far as destruction of human life is concerned, was minimal when compared to the atomic bombings of Hiroshima and Nagasaki. It grew in the 1970s as high quality, low cost goods began to enter this country from Japan and as they gained expertise in "our industry," semiconductors. It was further fueled by a

1983 Semiconductor Industry Association–sponsored report entitled "The Effect of Government Targeting on World Semiconductor Production." This document charged that government-induced reorganization of the semiconductor industry and direct government sponsorship of such joint industry efforts as the VLSI Cooperative Research Project (1975-1979) made it possible for the Japanese to effectively penetrate the U.S. market. Overlooked in such charges is the facts that much of the Japanese technology was eagerly licensed to them by U.S. semiconductor firms looking for quick financial return, and that the purchasers of the Japanese semiconductor products were some of our most prestigious and successful electronic original equipment manufacturers, including all the manufacturers of computers and data processing systems as well as the automotive manufacturers.

Despite such facts, the notion "Japan, Inc." is still popular in some circles. A recent book by Pat Choate titled *Agents of Influence* charges that the television industry in the United States was systematically destroyed by a cartel of Japanese manufacturers supported by the Japanese government and helped by a willing U.S. government under the influence of hundreds of American lobbyists and consultants. Wow! My own experience indicates that the Japanese manufacturers just did a better job at running their businesses. As previously noted, major manufacturers such as Zenith and RCA were quick to license their technology to Japanese television manufacturers and to turn to Japanese semiconductor companies, many also in the television business, for some of their critical IC requirements and thus directly helping their overseas competitors. However, it should be remembered that today there are more than 15,000 Americans employed in the television industry in this country by U.S. subsidiaries of foreign corporations, both Asian and European. Yet, it is strange that an entire industry that the U.S. created and one time dominated could ultimately be lost.

But while there is no monolithic "Japan, Inc." out to sabotage American industry, it is true that some Japanese companies have engaged in some questionable or even flatly illegal trade practices. In the case of the television industry, there was some dumping (that is, selling below manufacturing cost) of Japanese television sets in the U.S. market. Dumping is designed to drive competitors from the market; once the competitors are gone, the surviving companies are free to set prices at the levels they desire without having to worry about competitive pressures. A case history was given on the February 18, 1992 edition of *Frontline*, a Public Broadcasting System (PBS) television program. According to the *Frontline* investigative report (which, incidentally, was prepared cooperatively with a Japanese writer), in 1956 Matsushita formed an illegal cartel of Japanese television manufacturers. This cartel sold sets in the Japanese market at artificially high prices, and then dumped sets in the United States at prices so low that U.S. competitors were eventually forced out of the business. Higher prices on the invoices were shown to U.S. customs officials, while healthy rebates to retailers in the United States allowed them to offer cut-rate bargains to consumers. When this system was brought to light in 1968 by U.S. competitors, the U.S. government seemed benignly indifferent and collected only a fraction of the dumping duties originally levied. Recognizing that trade frictions were increasing, Matsushita bought Motorola's Quasar TV Division in the mid-1970s and all the Japanese competitors began to move to the United States for manufacturing (or at least assembly) to serve the American market. Today, the television set business is controlled by foreign corporations, both Japanese and European.

Sprague Electric experienced dumping firsthand, and was involved in several suits against Japanese capacitor manufacturers, including one over dipped solid tantalum capacitors, in the early 1970s. The most prominent of the Japanese firms charged was NEC. While dumping was definitely proven, the Tariff Commission (now the International Trade Commission, or ITC) ruled that no damage had been done to the U.S. industry. Sprague appealed, but no penalties were ever awarded. This was typical of the reaction of the United States government at the time. It was almost as if Japan was politically untouchable. So a few jobs, a company or two, maybe an industry were lost—so what? Did this help the Japanese firms penetrate the U.S. market? Yes. Were U.S. manufacturers injured? Yes, at least as far as dipped solids were concerned. However, Sprague and Kemet are still—even today—the dominant suppliers to the U.S. market, although both are experiencing tighter margins as the product mix shifts to surface mount devices.

The role of the Japanese government relative to industry has been changing and its influence waning as the competitiveness of business entities has strengthened. It was particularly heavy-handed following the war and between the mid-1960s and 1970s when it strictly controlled foreign investment. However, even then MITI was unable to rationalize the Japanese automotive industry as individual companies took action in relation to their own best interests and not those of the group. Even the true effectiveness of the much touted VLSI project mentioned previously has been questioned, at least relative to the initiatives of individual companies and the less well known NTT Project that supported DRAM development in the 1974-1980 period. There is no question that, unlike the United States, Japan has an industrial policy. It is based on the perception that the free market is flawed and must, to a degree, be controlled. These flaws are caused by poor information, the pursuit of short-term and individual goals by companies at the expense of the collective good, and lack of attention to national goals. In addition to MITI, there are other important government bureaucracies, including the Council for Science and Technology (STA) which develops strategic vision in science and technology, and the powerful Ministry of Education, which spends 50% of the government's research and development funds (versus 13% by MITI). However, it is MITI that appears to have the most influence. Today, more than anything else, it maintains an environment favorable to business and, through a number of techniques, signals where it feels the future technology and business thrusts should be. As tactics, it facilitates financial flows (including subsidies), attempts to influence the structure of industries under its jurisdiction, protects emerging industries, and attempts to regulate competition within selected sectors. MITI also helps reduce risk through subsidies and other financial vehicles, promotes exports, and negotiates trade issues. It also has been, and remains, the principle driving force behind the never ending drive within Japan for quality. Today, under pressure from the U.S. government, it is also trying desperately to increase U.S. imports. This is a tough task at best since corporations and the individual consumer are the final determinants in what is purchased from whom. Today, much of MITI's success comes from the respect it has within industry, the prestige that its programs carry, and because it employs some of the most competent and intelligent individuals in the country. (Wouldn't it be nice if the U.S. government had this image???)

By contrast, it is impossible to define any concise form of U.S. industrial policy or even reach consensus that one is necessary. If there is such a policy, it seems to be a belief in letting free market forces operate (up to a point) and to serve as the world's policeman. As pointed out by John A. Adam in "Competing in a Global Economy" in the April, 1990 issue of the IEEE *Spectrum*, the massive defense spending on research and development in this country has caused the Pentagon to become sort of a "ministry of technology" by default. In effect, by procurement policies and direct financial support, defense has become the principle targeted industry in the United States. Back in the 1950s and 1960s, such support did help foster the growth of much of the electronic components industry, including capacitors and semiconductors, as well as the computer and aerospace industries.

While many in the United States have long felt such defense spending necessary, it is increasingly being seen as one of the root causes of our problems in industrial competitiveness. As discussed earlier in this book, defense spending seems to be past the point of diminishing returns as far as being a source of "spillover" technologies for the industrial sector. And how safe is our defense structure if many of the semiconductors—and essentially all of the ceramic packages used to house military devices—and many of the other electronic components come from Japanese and Japanese-owned corporations? Sure, we have protected our critical IC and system designs; we just can't use them without foreign supply of these other crucial elements. Here is a clear case of where the loss of certain industries during the 1970s and 1980s is leaving the United States vulnerable to supply disruptions every bit as serious as the oil shortages of 1974 and 1979.

At the time this book was written, there was expectation in some quarters that the "peace dividend" resulting from reduced defense spending would help decrease the federal deficit and make us more competitive. Great care should be taken in drawing such a conclusion, at least over the short term. Defense spending has created an entire industry with specialized jobs that are not easily transferable to the industrial sector. Nor are defense contractors necessarily qualified to compete globally in commodity-driven consumer-based industries. Unfortunately, one of the first results of the "dividend" may well be an increase in unemployment.

Business/Government Relations

I believe in private enterprise and that the primary responsibility for success in the industrial field lies with each individual company. However, one of the very great strengths of Japan compared to the United States is not some conspiracy involving business and the government, but rather it is a stable political environment that strongly favors business. One of the ironies of the world today is that, after World War II, this cooperative environment in Japan grew out of the introduction of democratic principles by the occupying U.S. forces. Over the years it has been modified to be more consistent with a society and people who ethnically seek cooperation. In addition, some of the brightest and most competent Japanese citizens have chosen government service.

Such a relationship between government and business does not exist in the United States. Instead, we have a political system and politicians which tend to pit labor and management against each other rather than foster cooperation. Moreover, we often hear rhetoric from elected officials that infers—or flatly states—that, in effect, business is somehow bad. Most of our politicians have never had the experience of running a business. Many politicians seem to be unable to understand that profits are necessary not just for corporate survival;

they are the very reason we are able to provide jobs, invest in property, plant, and equipment, and afford the taxes and other expenses that are necessary to support education, social requirements, and the entire fabric of our society. Without successful businesses and the jobs they create, as we saw in the late 1980s and early 1990s in eastern Europe and the USSR, there is ultimately no society at all.

As a nation, cooperation does not come easily for us. We are such a pluralistic society that it is hard to get consensus on anything. Only at times when there is real crisis and our nation is threatened, as in time of war, do we all pull together. But I believe war is what we have now—an industrial war for survival and continuing improvement in our standard of living. Until, as a nation, we recognize and accept this, until business, government, and labor can reach an accommodation that creates more partnership than confrontation, we will continue to suffer a severe competitive disadvantage versus Japan and, ultimately, other emerging nations.

Everyone has his own agenda for the U.S. government. What is mine?

For a start, the cost of capital must be reduced. Unfortunately, no one is quite sure how to do it other than by reducing the federal deficit and increasing the savings rate. And a spirit of cooperation between government and industry must be created that stands the test of time, including changes in the White House and Congress. These sound like simplistic "motherhood and apple pie" statements, but they are true. *Any* steps to reduce the cost of capital and increase cooperation between business and government would be welcome.

Every effort should be made by the U.S. government to create an equal playing field and to penalize violations such as dumping, protectionism, and flagrant breaches of trade agreements. I also believe that the effort to reach at least 20% of the Japanese semiconductor market should continue. While results have been modest at best, the Japanese government and industry are certainly feeling the heat. However, if U.S. suppliers really want to play on that field, they must be willing and able to meet the stringent requirements of this demanding marketplace.

Strong government support of research and development should be continued with, however, a much greater emphasis on the industrial sector. This includes support not just of research but also of the commercialization process. Modest effort in this direction has been made through such initiatives as the Engineering Research Center (ERC) and State Industry University Cooperative Research Center (State IUCRC) programs sponsored by the National Science Foundation. I have served as a panelist on both and they are very interesting efforts, although the jury is still out on the results. Making research a "national priority," as recommended by the Council on Competitiveness, seems more like words than substance. Research already is a national priority; however, it has just been targeted too narrowly toward defense. The jury is still out on existing government-sponsored research consortia such as Sematech. From the results of Sematech so far, it seems that while little positive been accomplished, no real harm has been done. From my own observations, I think that such consortia are of very limited usefulness compared to the initiatives that must be taken by individual companies.

The need for strengthening the entire educational system, especially at the primary and secondary levels, is a must. But how to do so is not a matter of total dollars spent (unless you take into account the huge related social problems discussed earlier). As we noted earlier, the United States now spends a considerably higher percentage of its GNP on education (6.8%) than either Japan (5.0%) or Germany (4.5%). It is how the money is spent that is crucial.

No one can argue with the need for protecting intellectual property, especially in the U.S. which still leads the world in basic research. However, much of the problem lies in our legal system where cases are often settled only long after the damage has been done. Certainly the Japanese feel that intellectual property is important as they flood the world, including the United States, with patent applications.

When you get into tax policy, things really get murky. I am certainly no expert, but I believe reduction or elimination of the capital gains tax is appropriate, especially as a way of encouraging investment and causing investors to hold equities for much longer terms. This should allow management to take a longer term perspective, although your best companies already do within the context of the current tax structure. Still, much of the developed world does not tax capital gains, including Japan and Germany (who exempts long term gains only). In his book *Technological Competition in Global Industries,* David Methe' has shown that the availability of venture capital for investment in semiconductor start-ups increased in the United States during periods when the capital gains tax was lower (it has gone up and down periodically over the last two decades between 28% and 48%). However, investments were actually made only if technological innovation also existed. I would also recommend that the research and development tax credit be made permanent, at a rate of 20% or higher, and that depreciation of semiconductor equipment be accelerated.

If foreign companies want to be free to invest in the United States, then U.S. companies should be able to make investments outside the

United States. Unfortunately, it is almost impossible for U.S. companies to make such investments in Japan. In a recent telephone conversation with Lester Thurow of the Sloan School at MIT, I asked what he thought was the answer. (This was in relation to his recent comments in the *Boston Globe* mentioned earlier.) He said that the net result, if not solved, would be that Japanese companies would end up restricted on what they could buy and sell in this country. This is already the case in much of Europe. This problem will become even more severe in the future since, before the end of the decade, there will and must be a massive global consolidation of the electronics industry. There are just too many players for the business available. However, the buyers cannot only be Japanese. What also makes no sense is the fact that current anti-trust legislation in the U.S. makes it nearly impossible for strong U.S. competitors to merge for greater efficiency but allows foreign corporations to achieve the same result by buying U.S. competitors. The purchase of AVX by Kyocera is a perfect example. Our current anti-trust laws must be interpreted or modified to comprehend the realities of global competition, not just competition in the United States between U.S. corporations. A much more careful screening of foreign investments is needed to make sure that the ultimate result isn't destruction of U.S. competitiveness. This is exactly what has happened to the semiconductor infrastructure in the United States and what is underway in parts of the components industry as well. On the other hand, if U.S. corporations do not see fit to object or to offer viable alternatives to troubled companies seeking help, then such foreign investment is certainly better than closing the doors.

As far as federal standards related to things such as waste disposal, pollution, safety, and gas mileage are concerned, I feel they should be as stringent as reasonably possible. This is necessary for consumer protection, and, it should be noted, in most areas they already are just as tight or even tighter in Japan than in the United States.

At the time this book was being written, there was much concern in the United States over the rapid and steep increases in medical expenses in recent years. Medical care consumes a much higher percentage of the gross national product in the United States than it does in Japan and virtually all other industrialized nations. These increases have resulted in employers having to pay sharply higher medical insurance premiums to cover their employees, while employees are having to pay higher deductibles and having lower percentages of their total medical care costs covered by insurance. Every dollar that goes into escalating medical costs is one dollar less available for business investment or personal savings. I believe some sort of national health care plan to contain medical costs in the United States is a necessity.

The Cost of Capital

Everyone seems to agree that cost of capital is an important variable in industrial competitiveness and that it should be as low as possible, at least as far as corporate investing is concerned. On the other hand, it is not clear just how to compare it in different economies, unless one sticks to interest rates on relative government bonds as a starting point. The problem is further complicated by differing signals from different sources.

While historically interest rates have been considerably lower in Japan than in this country, recently they have been approaching U.S. levels. In a January 2, 1990 article in *The Harvard Business Review*, provocatively titled "Who Is Us," Robert Reich of the Kennedy School at Harvard stated that, due to the international movement of savings today, the cost of capital of different nations is converging. On the other hand, a 1990 report by the U.S. Office of Technology Assessment (OTA) estimated that the cost of capital in the U.S. is between 1 to 13% higher than in Japan. That is quite a spread!

There are two basic elements to cost of capital. The first is the interest rate of "riskless" financial instruments (in this country, the interest paid on U.S. Treasury notes) The second depends on the risk premium to be added to the government rate; this differs depending on the method of financing and the perceived risk of the investment by the lender. In the case of bank loans, this premium depends on the availability of capital, how the loan is secured, and how the bank views the borrower. In the United States, such borrowing has increasingly become more expensive and hard to get because of concerns with the viability of the overall economy, financial problems in many of the states, and problems in a number of lending institutions,

While there are many different corporate instruments, we will discuss only two, corporate bonds and common stock. Corporate bonds certainly are riskier than U.S. Treasury instruments, with the main risk being default. However, absent default, they guarantee interest payments and return of principle upon maturity. Both bank loans and corporate bonds do require the availability of cash flow to meet the required interest and principal payments. This is why the use of equity is sometimes referred to as "costless" since—unless there are dividend requirements—no cash flow is necessary to directly service the stockholder. On the other hand, equities are considered a higher risk instrument to the investor (due to the possibility of a total loss of the investment in the event of corporate insolvency) and therefore carry a higher risk premium than bonds. In this sense, they are the most expensive method of financing to the corporation, especially in the case of early stage financing such as venture capital. Here the primary assets of the company are often intellectual capital and goodwill which are difficult, if not impossible, to value. The market value of a stock represents what the stockholders, or owners, feel the corporation is worth, and the return they expect to receive through dividends, if any, and capital gains. If the stock falls relative to the market, this means that the owners are losing confidence and the company is vulnerable to takeover or acquisition. This is exactly what happened to Sprague Electric in 1976 when it was selling at well under book value and was acquired by General Cable Corporation (GCC)—later renamed GK Technologies—in one of the best deals GCC ever made.

In the United States, the interest rates on government securities are related to the government's need for money, its availability, and competition from other investment opportunities. The government's need for money is determined by its level of spending, as we explored earlier in this chapter. Interest rates also depend on whether the Federal Reserve is trying to stimulate the economy (lower rates) or control inflation (higher rates). This is where the deficit comes into play. In the early 1980s, forced to fund the shortfall between income and expense, the government was compelled to flood the market with securities. These were largely purchased by the Japanese who liked the high interest rates and who were flush with cash and had little place to put it. Thus Japan began to help finance the U.S. deficit and has been doing so ever since. "Riskless" U.S. government securities became riskier, at least to foreign investors, when devaluation of the dollar began toward the end of 1985. One of the major triggers to the October, 1987 stock market crash was a massive Japanese sell-off of U.S. government bonds. Since then, because of the weak dollar, they have begun to lean more toward investment in equities, real estate, and U.S. corporations.

So, if Japanese interest rates are lower than in the United States, they have an advantage in this part of the cost of capital equation. On the other hand, there is another major difference. Historically, the primary source of external financing in Japan has come from banks and from life insurance companies. For example, in the 1975–1978 period such sources represented 90% of the external financing for semiconductor companies. Although they have been rising the last several years, rates from such sources have historically been low. Why is this? Due to the high savings rate, there has been a ready availability of capital. Often the Japanese lender has an ownership position in the borrowing corporation. As long as the interest payments are made, there is little pressure on management as far as running the business is concerned. I observed this attitude first hand when I happened to be talking to one of the major lenders to Nichicon several years ago. As long as the interest payments on the bank loans were being met, they had no interest whatsoever in any involvement in the company's affairs, even though President Hirai of Nichicon was then over 70 years old.

In contrast, high technology companies in the U.S. depend predominately on equity as the primary source of external funds. As we have already discussed, this is the most expensive of all sources. So, whether due to lower interest rates, or to the risk premium assigned, or to both, logic would say that the cost of capital in Japan has been, and still is, lower than in the United States. Just how important is this? As one might expect, *very*. All other things being equal, cheaper money must be a competitive advantage. Certainly Marshall Butler, CEO of AVX, believes it is. In a recent letter he stated, "I really feel that the major difference between Japanese and American companies is their investment policies. (Because of a higher cost of capital) American companies require higher margins. The lower margins of Japanese companies free up money for lower pricing, more engineering, and more capital investments. The net result is a self-reinforcing spiral upward for the Japanese competitors and downward for the American competitors." That seems to sum it up pretty well.

The IC industry is a good example as its capital requirements began to outstrip the financial capabilities of all but the largest corporations or industrial groups. When Sprague first opened its IC plant in Worcester, Massachusetts in 1965, it made its own contact lithography printers at around $10,000 a copy. Today, the price of a stepper required for submicron design rules can run $1,000,000 or more. Different sources estimate the cost in 1993 of entry into the IC business at $750,000,000 if commodity devices such as DRAMs are the strategy. Unless the entire philosophy changes, by the end of the decade the price will certainly be in the billion dollar plus range. Even some of the second tier Japanese companies such as Oki are beginning to feel the crunch. These costs have created several new joint ventures and partnership arrangements. For example, Hewlett-Packard, Texas Instruments, Canon, and the Singapore Economic Board have created a joint-venture in Singapore to manufacture 4 M DRAMs. The total investment will be approximately $300,000,000. Texas Instruments is to be the sales arm with output allocated to the other investors in relation to their financial contribution. Texas Instruments has also joined with Acer, Inc. to invest in excess of $250,000,000 in Taiwan for a 4 M DRAM fab. In still another cooperative venture, IBM and Siemens have joined to build a 16 M DRAM wafer fab in France at a cost in excess of $600,000,000. Considered more controversial by industry insiders is the decision by Motorola and IBM to work together on integrating what is now a complex six-chip IC set used in workstations and PCs.

From these examples we can see not only the magnitude of the investment but also some of the different paths that are being taken. While some large companies appear to be prepared to go it alone, even big players such as IBM, Motorola, Siemens, Hewlett Packard, and Texas Instruments are looking to spread the risk by joint investments and joint ventures. In such a big money/big stakes game as ICs, it's clear that even incrementally less expensive capital for Japanese companies is a big advantage for them.

What can we do to change this "capital advantage" for the Japanese? To a degree, time is on our side as the interest rates of the two countries converge. The strongest action for the United States would obviously be a concerted and lasting effort by the government to balance the budget. (Based on the complete lack of success in doing so up to now, however, I see little chance of this happening in the foreseeable future barring a world-wide financial disaster.) Increasing the personal savings rate would also help as far as bank financing is concerned, as would lowering or eliminating taxes on long term capital gains on appreciated equities. I see no way, however, of getting a savings rate anywhere near that in Japan unless we cut our social security programs to the lower levels found in Japan (see Chapter 4).

I should point out that not everyone agrees the Japanese currently enjoy much of an advantage in the cost of money. For example, in "The U.S. Has a New Weapon: Low-Cost Capital," in the July 29, 1991 issue of *Business Week*, Christopher Farrell describes the convergence of capital costs around the world and the United States' apparent growing advantage in this important factor. (His conclusion must be considered with caution since this same article also talks of the improving U.S. economy in 1991, a prediction that proved to be much too optimistic.)

There is an important reason that will dramatically increase the cost of capital in Japan over the next several years, however. Much of the financing by Japanese corporations the last several years has been done through equity-warrant bonds or a close equivalent, low-coupon convertible bonds. In both cases the interest rate of the bonds themselves is extremely low, say 1-3% (or even less). The equity-warrant bonds carry a warrant that is convertible to common stock in the issuing firm at a modest premium over the stock price at the time of the issue. With a rising Nikkei stock market, investors' use of the warrants to purchase common stock allowed redemption of the bond upon maturity at zero to very low cost. The rub is that, since many of these securities were issued, the Japanese stock market has fallen dramatically and the warrants are essentially worthless. Without conversion, the original bonds can only be paid-off upon maturity by much more expensive refinancing. It is estimated that there is some $90,000,000,000 of such borrowing which will come due the next several years and which will require refinancing rates closer to 6.5% versus the originally planned 1.5%. This will dramatically raise the cost of capital in Japan and slow down investment over the next few years both in local capital spending and overseas. This important development was covered in more detail by Gale Eisenstodt in "Useless Warrants" in the November 25, 1991 issue of *Forbes.*

In summary, in the absence of definitive comparative data, the cost of capital in Japan has historically been lower than in the U.S. Even should interest rates become comparable, the highly leveraged position of most Japanese firms, along with the often close relationship between lender and borrower, should continue to favor Japan. However, maturing of the equity-warrant bonds we have described should dramatically increase the cost of capital and inhibit the future ability of Japanese corporations to use such cheap financing schemes. A protracted business down-turn in Japan would seriously jeopardize the financial stability of Japan's highly leveraged corporations.

Actions required by the U.S. government to help further close the cost of capital gap include reduction in the deficit and tax policies that favor both personal savings and long term retention of equities. Of these, the deficit is certainly the most intractable.

CHAPTER 11

Challenges to Japan

Japan and its people have been subject to a great deal of stereotyping throughout history. In World War II, the stereotyping was highly racist, depicting the Japanese as bloodthirsty, barely human beasts. Today, the stereotyping of the Japanese is the direct opposite—a completely unified industrial utopia populated by superhumans. The only thing this new stereotype has in common with the older one is that both are equally wrong and misleading.

It was true as these words were being written in mid-1992 that things looked very favorable for the Japanese in electronics. We have already examined their formidable strengths and remarkable achievements earlier in this book. But nothing is ever permanent. Thirty years ago, the notion that Japan would be overwhelming the United States in such industries as electronics or automobiles would have been considered laughable. The Soviet Union was seen as the main threat to the United States back then. There was a lot of serious talk three decades ago that the impressive Soviet space achievements (including the first satellite in 1957 and the first man in space in 1961) indicated Soviet technological supremacy over the United States. Such technological superiority, many felt, would be translated into military and economic superiority and the eventual world triumph of communism.

No one laughs at the Japanese electronics and automobile industries these days, and the Soviet Union no longer exists. In retrospect, a lot of "experts" of the 1960s look incredibly foolish. This is a reminder of an important point we must keep in mind: things change. Japan's current leadership in several key industries is not an immutable fact of nature or a necessarily permanent condition. Japan is facing several internal and external challenges in the years ahead, and whether Japan can maintain its current leadership role will depend on its response to those challenges. In this chapter, we will examine challenges posed by changes within Japan itself as well as the rise of competing nations in Asia and the European Community.

The Other Japan

The current image of Japan portrayed in much of the American media is of a nation of well educated, hard driving people who love their country and their work, who believe they can do things better than anyone else, and who are bent on economic domination of the rest of the world. Protected by lifetime employment and unencumbered by either the need for a military presence or the social problems facing many other nations, they concentrate on refining those parts of the manufacturing process that create competitive advantage in the industrial world. Believing themselves ethnically superior and with a government that maintains the type of environment within which industry flourishes, an energetic and driven work force rushes off to work each day bent on doing things better than the next competitor, be they Japanese or foreign. Within the protective shell of massive industrial teams, economic juggernauts assault and overcome ever more stringent goals and standards. As a result of this, perfect products flood the world and within colossal metropolises, the financial resources that flow from this onslaught threaten to economically destabilize the rest of the world.

Who can withstand this inexorable force? Let's all become "Japanese"!

But wait a minute. Let's stop and look beyond what we see on the surface and what we read.

Much of what we have described in caricature is true, as far as it goes. But let's remember some of the points we have previously noted in this book. Lifetime employment covers much less than half of the work force—that is, male employees of large corporations—and this benefit may disappear in time for them as well. Women are still second-rate citizens, mandatory retirement is now 60 (only recently up from 55), and life after retirement is seldom very easy. The safety net that covers so much of the U.S. population in the form of Social Security, pension programs, and medical care is much thinner in Japan. This is one of the principal reasons for the large savings rate. The other is limited options on where to spend these funds.

Yes, the Japanese work very hard, partly because of their cultural background and partly because of limited alternatives. A foreigner is seldom asked to visit a Japanese home. The first time I did so was at the invitation of Archie Suzuki (yes, that is really what we called him) who headed our Tokyo sales office during part of the 1980s. It was a long trip there by subway and train, well over an hour. However, upon arrival we were in a congested extension of the city and not the rolling hills and open spaces of, for example, Westchester County. The house was two stories high and tiny. The downstairs consisted of a small kitchen and a combination living room, study, and TV room that was filled with consumer electronic equipment. I also remember an upright piano and a narrow staircase that disappeared into a dark second floor. In addition to Archie, his wife, and two children, two elderly people lived there as well. I believe they were Archie's parents and I still have no idea where they all slept. It was a very pleasant evening and I was honored to be there. When I left for the long trip back to Tokyo I also noticed that there was no land whatsoever around the structure, and barely space enough to squeeze in a carport and a couple of bicycles. Proud as Archie was of his home and family, and happy as he was to have me there, I can also see why he spent so much time at work—there was no room for him at home. Visits to other middle class Japanese homes were very similar. Although I am not sure how you measure "standard of living," I came away with a new understanding of how very lucky we are in the United States.

Japan has a rapidly aging population and is facing potentially explosive problems because of this. One of the major reasons behind mechanization and offshore manufacturing is to compensate for a flat or even decreasing labor pool. More of the best and brightest young people are shunning the engineering disciplines for—God forbid!—financial services or other less demanding specialties. This is leading to shortages in critical fields. Another difficult issue is how to care for the aged. The Japanese are living longer than before. Nursing home care is much rarer in Japan and the children have traditionally supported their aging parents. This is becoming harder, however, as more women continue to work after marriage and children. Recently one of my Japanese friends complained bitterly that he just didn't understand why his wife was unwilling to take care of his parents anymore since "that is how it always has been done."

Because of an increasingly tight labor pool, job hopping is becoming more common. In addition, as affluence continues to grow and the Japanese travel and see more of the rest of the world, a number of other changes are occurring. There is an increasing aversion to the manufacturing floor, up to now one of the main strengths of the country. Although most engineering graduates still head toward industry, this skill pool is also tightening. More illegal aliens and women are beginning to find their way into the work force, and, in the case of the latter, not just at the operator level. Unhappy consumers are beginning to realize that they are directly financing the largely closed Japanese market through sky high prices on many staples and higher domestic than export prices on such luxury items as cameras and autos. In addition, land and housing prices are now so high that they are largely out of reach of all but the most affluent. Much of Japan's wealth is a direct product of the scarcity of land. Owning a home similar to that of a typical middle class American is enough to make one a millionaire (at least on paper) in Tokyo proper.

When we think of Japan, we generally think of the great metropolises such as Tokyo, Osaka, and Nagoya. This is not surprising since almost half of the entire population lives in and around these three great cities, and more than 50% of all corporate income is created in Tokyo. The "flip side" is a land of decaying farmland, crumbling fishing ports, and sunset industries. Yes, not everything is high tech in Japan. Problem industries include agriculture, petroleum, aluminum smelting, food, paper, and chemicals. Nor are all products perfect. While anything but an extensive survey, I have had nothing but problems with my Pentax camera, two different models of Panasonic answering machines have had malfunctions (one of them fatal), and an assortment of different Japanese televisions and stereo systems have had random defects, much like their Zenith, RCA, and Magnavox predecessors. Perhaps most frustrating has been a series of wonderful Casio sport watches. Their replaceable battery life is between three and five years, depending on the model; however, the plastic straps invariably break in less than two. Try as I have, I have yet to find a replacement that fits. As the result my bureau top is littered with such watches, "ticking" happily away and useless (that is unless you want to carry your wristwatch in your pocket).

While I never saw the type of extreme poverty we have in some parts of the U.S., there are homeless in Japan and you find them in many of the same places as you do here: in train stations, under bridges, wherever they can find protection from the elements. They also can be dangerous. While staying at the Keio Plaza Hotel in Tokyo and questioning what might be a good route to jog, I was told to stay away from several nearby parks because they were filled with indigents and it was unsafe to go there. It sounded just like Central Park in New York City.

Bankruptcies do happen in Japan, executives get fired, and scandals do occur, as those that have recently rocked NTT (the national telephone company) and the securities industry. A cover article, "Hidden Japan," by Robert Neff and several co-authors in the August 26, 1991 issue of *Business Week* highlights the cozy relationships that are part of doing business in Japan and that exist between politicians, business executives, bureaucrats, and even gangsters. These came to light as part of the Nomura Securities scandal. Since the Japanese system is based more on personal relationships than on objectively following the letter of the law, such embarrassing exposes may not seem particularly corrupt to the average Japanese. However, they do to the Western world. In addition, if that is the way business is done, it represents one more way outsiders are prevented from direct access to the Japanese market and gives the impression to the outside world that there definitely is some sort of conspiracy going on. As Japanese corporations become more international, they will have to change "business as usual" to conform more realistically with the rest of the world.

While in the past many Japanese manufacturers have played the competitive game using a different set of rules, they are also like us in more ways than we think. Certainly the U.S. has had more than its share of scandals and collusion between business and government officials. However, the fact that both nations demonstrate human frailties should give us little solace. There is no question that, as the Japanese economy matures and the nation becomes increasingly wealth driven, more of the problems we have just discussed will arise and worsen. To a large degree this is what happened in the U.S. in the glory years following the end of World War II. Whether or not the strong ethnic commonalty of the Japanese people will slow or even stop this process remains to be seen. Whatever happens, the United States cannot sit around and wait for their fall. Following World War II, the Japanese certainly didn't sit around waiting for American power to decline.

Japanese politics are also in a period of transition and strain. Apportionment for the Diet (the Japanese legislature) has not been changed as rapidly as population shifts, resulting in rural areas having a disproportionate share of seats in the Diet and influence. This explains, for example, the restrictions on rice imports from the United States and the apparent willingness of the Japanese government to tolerate quotas on its automotive and electronics exports in order to protect Japanese rice farmers.

What about the Japanese as individuals? As in the United States, it is very hard to generalize. As you can probably guess by now, I like the Japanese and have a number I consider friends. I must admit that the increasing arrogance I see in some successful CEOs is becoming annoying. On the other hand, in general the middle managers are about as fine a group of individuals as you will ever find. They seem to have a great deal of humility. Instead of talking about how good they are, they just go ahead and do their jobs—and very well indeed, thank you. Still, there are some disturbing trends with this group and even blue collar workers as well. Recent articles have noted that the "ugly American" is being replaced by the "ugly Japanese" when it comes to arrogant, overbearing behavior while abroad. Unfortunately, insensitive and insular behavior makes "Japan bashing" easy and readily believed.

"Japan bashing" is also inflamed by the shocking racism displayed by some Japanese public officials. In the early 1990s, there were statements by Japanese officials such as attributing the lower average IQ scores in the United States to the presence of blacks and Latinos in the American average. Statements critical of American workers by some Japanese officials also have had racist overtones. While the Japanese officials making such statements invariably retract them after criticism from the United States, the retractions and apologies often seem perfunctory and insincere; some of the speakers seem genuinely puzzled why their statements are considered offensive. Within Asia, Japan is criticized by China and Korea for the treatment that Japanese citizens of Korean or Chinese descent receive in Japan. (And China and Korea have not forgotten the Japanese atrocities of World War II, driven in large measure by Japan's sense of racial superiority.)

These attitudes are not restricted to government officials. A recent issue of *Technical Communication*, the journal of the Society for Technical Communication, carried a report on Americans who work in Japan as writers and editors of English technical manuals. Several of those Americans reported astounding incidents of racism in their workplaces. For example, one woman wrote that she was approached by some members of the engineering staff of her company for help in a "technical test." Bewildered but eager to help, she consented. The "test" consisted of trying to read something displayed on a color monitor. The woman had blue eyes, and there had apparently been a considerable argument among the engineering staff as to whether it was possible for someone with blue eyes to discern certain colors!

Notions of the racial purity and "specialness" of Japan have long been a part of Japanese culture and national identity. In World War II, such attitudes were explicit. After Japan's defeat, those attitudes were buried. Today, they are regaining credence among some right-wing nationalists. As we saw at the beginning of Chapter 6, even some Japanese businessmen are starting to express opinions that seem drenched with the notion of Japanese superiority and specialness. It is likely that these racial attitudes will become an increasing source of conflict and trouble for Japan as it becomes an important political player on the international scene.

Like people everywhere, the Japanese complain about and fret over their young people. But it is true that Japanese youth are beginning to question many of the pillars upon which much of the post-World War II economic miracle was based, particularly unquestioned loyalty to the corporation and long working hours. These same young people also wonder why the average Japanese standard of living is still so low compared to the United States and much of Europe if Japan is so wealthy as a nation. While it is certain that the vast majority of these young people will eventually join the Japanese "establishment" (much like America's hippies of the 1960s turned into the yuppies of the 1980s), there is clearly a dramatic difference in attitudes between the generations in Japan.

In a review of the first draft of the manuscript for this book, one reviewer remarked that this section seemed like "Japan bashing." Unless one believes that any criticism of Japan is automatically "Japan bashing," I don't think that remark is valid. The point of this section is to remind ourselves that the Japanese are human and have both good and bad points—just like us. They are not villains, nor are they supermen. Like us, they have problems that urgently demand solution.

I believe that as Japanese corporations become more global, they will be forced to play by international rules. If they can't or won't, they will ultimately be prevented from playing the game at all. We can certainly learn much from the Japanese, and in turn the Japanese can learn much from the United States.

"The Tigers" and "The Cubs"

As far as growth is concerned, the Western Pacific is currently where the action is. This sphere includes Japan and the so-called newly industrialized countries (NICs): South Korea, Taiwan, Hong Kong, and Singapore. These are collectively known as the "Four Tigers." Member countries of the Association of South East Asian Nations (ASEAN) are often known as the "tiger cubs." These nations include Malaysia, Thailand, the Philippines, and Indonesia. Including the Chinese coastal provinces of Guangdong and Fujian, the region holds close to 600,000,000 people and has a combined GNP close to $4,000,000,000,000. By the year 2000, this region's GNP is forecast to be greater than that of western Europe and equal to the United States. Will this region also emerge as a major competitor to Japan, perhaps beating the Japanese at their own game?

TABLE 11-1

Key Statistics on Western Pacific Economies
(in billions of dollars)

Country	*GNP*	*Engineering Work Force*	*Major Electronics Companies*
Japan	$2927	1,124,000 (1985)	Eleven giant companies with electronic sales over $5, ranging from Ricoh ($5) to Matsushita ($31)
The "Four Tigers":			
South Korea	$240	18,000 (1985)	Four huge diversified groups or "chaebols" (Samsung, Hyundai, Lucky-Goldstar, and Darwoo) with decreasing revenues from $43 to $16.
Taiwan	$164	180,000 (1990)	Tatung ($1.15)
Hong Kong	$71	42,000 (1990)	Video Technology International Holdings ($0.3)
Singapore	$34	10,000 (1991)	Singapore Technologies Industrial Corp. ($0.35)
"Tiger Cubs":			
Malaysia	$42	—	—
Thailand	$80	—	—
Philippines	$45	—	—
Indonesia	$96	—	—

Table 11-1 gives some comparative data about this region, including Japan.

As one can quickly determine from Table 11-1, Japan remains overwhelmingly dominant in all these countries in terms of economic power, engineering skills, and diversified large corporations despite the tremendous strides made by the others in recent years. Only the four huge South Korean conglomerates come even close to the Japanese, at least as far as economic leverage is concerned. As a matter of fact, through investments throughout the region, Japanese corporations are in the process of integrating many of the skills of the other countries—be they technical or low cost labor—into their manufacturing systems. In the process they are also gaining early access to a potentially huge market.

According to Andrew Tanzer in his article "What's Wrong With This Picture?" in the November 26, 1990 issue of *Forbes,* this all came about as the result of the "Plaza Accord" which then-Treasury Secretary James Baker arranged in September of 1985. This accord devalued the dollar and, as a result, over time the yen strengthened versus the dollar from 240 ¥ to the dollar to about 130¥. Due to the resulting higher comparative labor costs, the Japanese were forced to quickly move high labor content production to such places as South Korea, Taiwan, and Singapore. More recently, they have been investing heavily in manufacturing in Malaysia, Thailand, the Philippines, and Indonesia. The ever tightening labor market in Japan has also made this a necessity. By the early 1990s, on an *annual* basis Japanese companies had become the largest investors in essentially every country in the region, well ahead of U.S. corporations or any other western competitors. As they concentrate more on high technology at home, they are willing to send mature consumer-oriented industries such as small TVs, radios, and cameras to other Asian countries. Of the East Asian automotive market, both through exports and local assembly plants, Japanese corporations currently control an overwhelming 80%.

South Korea is known primarily for consumer electronics and produces 15% of the world's color TV sets and VCRs, as well as 25% of the microwave ovens. It also produces low-end automobiles, of which the best known is the Hyundai line. One company, Pohang Iron and Steel or "Posco," is the third largest steelmaker in the world. In addition, there is a high tech push. In ICs, Korean firms such as Hyundai and Goldstar, both recent entries in the increasingly crowded 4 M DRAM fray, account for 3% of the total worldwide IC production and 15% of the DRAMs. This capability is, to me, surprising, in light of the very low number of engineers in the work force shown in Table 11-1. My guess is this number has increased dramatically since the 1985, the year covered in the table for Korea. Through 1991, U.S. cumulative investment in Korea still led that of Japan by a factor of 3 to 2, due, no doubt, to our large military presence.

Taiwan is also known for consumer electronics, although, unlike South Korea, there is only one major electronics corporation, Tatung. The country is also involved in the semiconductor market and provides roughly 1% of the world IC production, again primarily in DRAMs. Hong Kong has no large electronic corporations. Rather there are a number of small firms that concentrate on providing electronic components and low-end consumer products such as television receivers, radios, and audio recorders. It also serves as the region's financial capital and economic window into mainland China. At the time this book was written, professionals continued to exit Hong Kong, fearing what will happen when the colony comes under China's jurisdiction in 1997. Many I know and have talked to come from families that originally fled the mainland in the late 1940s and want no part of what they escaped. My own guess is that the Chinese will be very careful not to upset the important export and industrial door that Hong Kong provides to the outside world.

Singapore is in a class by itself. Tiny—a population of less than 3,000,000—it still has a per capita income second only to Japan in the region. It is a beautiful place, although I feel that "the Paris of Asia" is an extremely misleading descriptor. It is lush, distinctly Asian, and has been successful in attracting foreign investment in higher end industries such as computers and telecommunications. U.S. automotive companies such as General Motors also do electronic assembly there. An important additional function of Singapore is to serve as an interface between Malaysia and the outside world.

Sprague Electric has had a manufacturing presence off and on over the years in many of the East Asian nations, including Malaysia, Taiwan, the Philippines, Hong Kong, and, on a subcontract basis, South Korea. There has also been a sales function in all these countries, as well as in Singapore and Australia. For many years, in addition to manufacturing, Hong Kong has also served as Sprague's Asian headquarters. To a large extent, most manufacturing has been to take advantage of low labor costs rather than local market penetration. By far the most consistently successful of these has been the Philippines, where the Semiconductor Group has done back-end assembly, and more recently test, since the 1970s. Despite the continuing political turmoil in the country, this success has been primarily due the extraordinary skills of an American expatriate, Fred Reiersen, managing director of Sprague Philippines, Inc. Fred has been able to work through the minefield of local interests and, with excellent support from the U.S., create one of the most efficient and highly productive operations of its kind anywhere in Asia. Since the Filipino labor force is not known as the most the region's most capable, this shows what tight screening, excellent training, first class supervision, and leadership can accomplish. I doubt that Fred will ever return to the United States for good, preferring to be, as I call him, "king of the Philippines." (Excuse me, Fred; I know you don't like that title.)

How do these nations rank? Next to Japan—as far as economic muscle, major industrial groups, economic planning, work ethic, an emphasis on education, and literacy are concerned—South Korea is number two in the region. It also has an excellent infrastructure of roads and communications, a strong export bent, and strict import barriers. Besides being purely ethnic in origin, the deep-seated desire to succeed of the Koreans is driven by a hatred for Japan that results from the brutalization and degradation of the country and its people that occurred while it was a Japanese colony from 1910 until after World War II. Many Americans seem to equate Japan and Korea, but the differences between them are deep and profound. Korea also has a very different "feel," for want of a better word, something I clearly sensed during a mid-December visit I made in the early 1980s. The purpose was to investigate the possibility of a capacitor joint venture or other relationship with one or more potential partners.

Seoul is a cold place in December. In fact, one has the feeling that it is cold all the time, although this is certainly not true during the summertime. As one drives in from the airport, the city seems to rise out of the snow and rocky landscape like some newly created phoenix. I stayed at an excellent, high rise hotel that rose above the city and surprised me with fresh orange juice to accompany the "Today Show" at its usual morning time on television (on a delayed basis, of course). The first morning was spent getting acquainted with our hosts and unimpressive plant visits to local manufacturers. After a forgettable lunch, we climbed into several large limousines for a much more memorable trip to an aluminum capacitor operation several hours from Seoul. I was struck by the barrenness of the countryside and the fact that it was completely devoid of large trees. Upon questioning this, a bitter reply said this was because the Japanese had completely denuded the country of timber during the latter stages of the occupation.

When we finally reached the factory, set near a small town in a valley, it was a study in contrasts. It turned out that the technology and equipment had been licensed and purchased from a Japanese partner over a period of years. As the result, some was very current and modern while other lines were equipped with older generation gear. As a result, the manufacturing layout was a hodge-podge and, overall, in no way competitive with what I had seen in equivalent Japanese factories. However, there was no question about the work ethic or labor productivity. Not an operator even glanced at us as we toured the lines. Management also had an interesting approach to insuring productivity. While the small, spartan, management area and manufacturing spaces were well heated and comfortable, the hallways and toilets were completely unheated. The latter were just as uncomfortable as a skiers outhouse atop Mt. Tremblant north of Montreal in winter. As a result, except for emergencies, no one ever left their work stations except for planned breaks. I never could figure out where those were held.

Meeting in a small conference room after our tour, Don McGuiness, who was accompanying me on this trip, and I questioned how they were planning to set up manufacturing for us. It was unclear how this could be done since the technology had been licensed from one of our Japanese competitors. Not to worry—they would merely duplicate the latest equipment and transfer the best operators, technicians, and management to the new Sprague operation. When I questioned the ethics of this, the reply was the same. Not to worry—that was their problem. I left for the return to the city with a distinctly uneasy feeling. Just what kind of people were we dealing with?

That evening, Don and I plus three of our hosts went to my first—and hopefully last—*keising* party. Somewhat like the business clubs in Kyoto mentioned earlier, the idea was to create male bonding between potential business partners. But what a difference! There were three lovely young ladies in attendance, and my "date" was a tiny young thing called "Miss Moon." She was not so young, as it turned out, since this was only part time work and she was also married. We sat around a low table in a relatively small room and Don and I each had a sort of king's throne against the wall. As the senior American, mine was somewhat larger than his and in the center of the table. The girls fed us from a seemingly endless number of ever changing dishes. I remember at least three different soups, an equal number of desserts spread out through the meal, and some meat dishes I was wise enough not to question. We also played such silly games as passing a grapefruit under our chins around the table. While this was a great way to get to know "Miss Moon," woe be to anyone who dropped the fruit. This required downing the equivalent of a jigger of dark, resinous rice wine at one gulp. After several of those the evening would undoubtedly come to a sudden and unhappy end. Don and I did pretty well, but "Joe," one of our Korean hosts, was soon beet red, aggressive, and not long for this world.

Just as this bonding exercise began to pale, an orchestra (if you could call it that,) appeared and I had the chance to dance with "Miss Moon." Lovely as her face was, due to the layers of bright silks that covered her there was no way to tell what her body was like except that it was small. Just as I began to feel this was more like it, at around 11:00 PM a bell rang, the orchestra packed up, and my partner stepped back, pointed her little finger at my face, and said, "you go home now." Suddenly Don and I were out in the street, which was so empty that it seemed like there must be a curfew, and completely on our own to find a way back to the hotel. So much for Korean business hospitality. We did finally hail a cab and returned safely. As I got out of the elevator on my floor, I noticed that there was a "concierge" at a desk checking each guest as they arrived. She looked more like a guard at the entrance to a prison cell block.

There is little else to say about the remainder of our Korean visit except that we did not consummate an aluminum capacitor deal, although we did agree to subcontract some MLC assembly and test. This also fell through when we later decided to deemphasize and eventually exit this product area.

The point of my little tale is that Korea is not Japan, nor are the other "tigers" or "cubs." Each nation has its own distinct social, political, and business culture, and what "works" or is "normal" in one nation is not necessarily the same in others. Each nation will follow its own path in development and must be viewed individually.

At the end of the 1980s, total cumulative U.S. investment in the "four tigers" still exceeded that of Japan. By now, I believe this is no longer true, especially if you include the ASEAN group. Japanese companies are investing everywhere, both to extend their manufacturing base and to control the Asian market. Despite a World War II carryover of deep dislike the other nations in the region have for the Japanese, they seem perfectly willing to accept the ready cash and improved economies Japanese investment brings. As Japan expands, it will be increasingly hard for U.S. and European companies to remain competitive in the Western Pacific.

While at first glance South Korea appears poised to challenge Japan in certain mature, lower technology electronic areas such as consumer, overall it is highly doubtful they will ever come even close to catching up. The truth is they are just too far behind. The only nation with a chance of remaining even is the United States, especially those firms such as IBM who already are well established in Japan itself. For all but a few others, I believe time is running out.

Europe

The European Community consists of Germany, France, Italy, Britain, Spain, the Netherlands, Denmark, Belgium, Greece, Portugal, Ireland, and Luxembourg. The combined economies of these countries include some 340,000,000 people with a GNP close to $6,000,000,000,000. Austria, Finland, Iceland, Norway, Sweden, Switzerland, and Liechtenstein participate in some of the EC initiatives and add another 40,000,000 people. By far the most important members from an economic standpoint are Germany, France, Italy, and Britain.

What kind of challenge does Europe pose to Japan?

"EC 1992," the creation of "United States of Europe," has been hailed by some as the most important political and economic event of this century. It is as if this huge market opportunity suddenly appeared out of nowhere, ripe for the picking. This is a silly belief since the market has always been there, although over time implementation of the scheduled initiatives are expected to improve the combined economies by close to 5%, or $250,000,000,000. While this is nothing to be sneezed at, it may well also be optimistic. In addition, a number of foreign firms have been successful in penetrating the European market for many years, including IBM, GM, Ford, DEC, Hewlett-Packard, and a host of others.

An important concern is whether or not what is really being created is "Fortress Europe." This is thought of as a huge, impregnable combined economic sphere that not only keeps foreign corporations out but also within which the member countries work together to overcome the competitive advantages currently enjoyed by U.S. and Japanese corporations. While 1992 certainly wasn't designed to help outsiders, that may well be the end result. The big losers may be many of the European corporations which it was meant to help.

The European Community to be created by December 31, 1992 consists of literally hundreds of different elements. The most important include "open borders" for much easier movement of people, goods, and services within the Community, common technical and other standards, harmonization of different tax policies (including a common value added tax, or VAT), and a single central bank and common currency, known as the European Currency Unit, or ECU. There will be changes in the labor laws to allow professional persons (such as a doctor) trained in one country to work anywhere within the Community and open bidding on public projects throughout the EC. As one might anticipate, it will take time for all of these to be implemented and some, such as tax policy and common currency, may not be in place until early in the next century. (Moreover—as the currency crisis of September, 1992 demonstrated—it also remains to be seen if the individual nations of the Community can really sublimate their national interests to those of the Community over the long term.)

The planned benefits of EC 1992 are many. It will be easier for member corporations to operate within the Community. It should also make them more efficient as they combine duplicate facilities once felt necessary to serve different countries. This will also help foreign firms that are already well established in the EC, and those that make the effort to become so. Corporations that do not have local presence will find business much more difficult due to rules of origin and local content requirements. While EC 1992 is supposed to lead to more cooperative research and development by member companies, little positive had occured at the time this book was written. (In fact, it is unclear just how it will make EC corporations more competitive in high technology areas such as ICs and computers which are already dominated by U.S. and Japanese competitors.) Finally, a European-wide telecommunications infrastructure is planned, although full implementation will take years.

There are also a large number of issues that must be resolved, some of which are highly intractable. For example, it is unclear just how "open" the EC will be, either to members or to outsiders. The telecommunications network consolidation is enormously complex and will take years to implement. This is also true of rationalization of all the different standards that currently exist within different countries. And the entire area of tax policy and common currency is a minefield of different interests.

Elimination of government subsidies that have existed in different countries in such areas as steel, aircraft, and agriculture will not be easily implemented, especially if they are required for survival of a local industry. The entire problem of rules of origin and local content has already created difficulties between the different interests of key Community members. (We will also have more to say about this shortly.) In many cases goods that are free to flow between EC members still must be personalized for local consumption. For example, the British drive on the left side of the road and prefer to have washing machines that are top-loading. The French drive on the right and front-load their washers and dryers.

Different countries also have different requirements in such diverse areas as environmental issues, social action programs, and legal directives. Special interests will complicate rationalization of all of these.

What are U.S. corporations doing to respond to the new Europe that is being created? Established companies are working very hard to further consolidate the positions they have taken years to establish. This includes expanded investment, expansion, and increasing cooperative ventures with EC member companies. Cumulative U.S. investment of $150,000,000,000 in Europe through the end of 1989 still far exceeds that of any other country, including Japan. Led by IBM, U.S. corporations dominate the computer market. Nor are they standing still. New entries such as Apple and Compaq are establishing beachheads. "Big Blue" is running faster than anyone else. For example, IBM signed a joint development agreement with Germany's Siemens to develop the next generation 64 M DRAM at IBM's facility in East Fishkill, New York. Unlike Japan, U.S. automakers such as General Motors and Ford have long-established and profitable operations in Europe.

Pete Maden of Sprague see little change for his Tours, France solid tantalum capacitor plant resulting from 1992. Over the years Sprague Electric has had factories in Scotland, West Germany, France, and Belgium. While only the last two remain, they were all originally created to primarily serve U.S. multinationals in Europe such as IBM. Sprague also learned to sell product manufactured in one country to major corporations throughout Europe, both U.S. multinationals and local. While EC 1992 will make this easier, it is unclear how much additional business it will create.

Maden also feels that nationalistic feelings will die hard and that, if local industries exist in a particular commodity, local consumers will continue to buy them if at all competitive. According to Pete, the biggest winners long term will probably be the Japanese. So many individual communities throughout Europe are depressed and need investment that, regardless of national feelings, if Japanese corporations come with money to spend and jobs to create, they will be allowed to do so.

The Japanese are everywhere in Europe with capital to invest. Much of their original base has been in the U.K. which has welcomed them with open arms, especially during the Thatcher administration. To date, close to 40% of Japanese EC investment has been in Britain, much of it in auto assembly plants for giants such as Nissan, Toyota, and Honda. They are also expanding in semiconductors and in computers. For example, in a move that shook much of the EC community, Fujitsu acquired the British computer manufacturer ICL in 1990. By the year 2000, it is estimated that close to 16% of the British labor force will be working for Japanese owners.

The Japanese are now moving rapidly into Germany, due to the size of the market and future potential due to reunification. In the early 1990s, Germany purchased 40% of all Japanese exports to the EC and, in turn, is the largest European exporter to Japan. Cooperative ventures are also binding German and Japanese corporations, including the recent broad agreement between the Mitsubishi Group and Daimler-Benz covering research and development and manufacturing in such important areas as automobiles, aerospace, microelectronics, and services. Mitsubishi Electric is also building an IC plant in the depressed coal region of Aachen, Germany.

Of great concern to both U.S. and European semiconductor manufacturers is the massive investment throughout Europe by Japanese IC giants, including Mitsubishi, Fujitsu, and Hitachi. The European position is particularly desperate in both ICs and computers. In 1991, the European electronics trade deficit was $35,000,000,000 despite a reasonably healthy consumer electronics industry (especially in television for the local market). There have been massive restructurings and layoffs at NV Philips of the Netherlands, the joint Italian-French corporation SGS-Thomson, and Germany's Siemens AG. Despite additional captive semiconductor sales at Philips and Siemens, none of these three is really an important world factor and, on a global basis, rank only tenth, twelfth, and fifteenth, respectively, in the world semiconductor market. There has even been discussion of a possible merger between SGS-Thomson and Siemens, although this was placed on hold due to the massive related restructuring costs.

The EC has taken action to slowdown the Japanese in ICs by first levying 60% duties on the imports of eleven DRAM manufacturers due to unfair pricing or dumping in the European market, and then modification of this to quarterly set floor pricing on such imports. Much to the chagrin of U.S. and Japanese exporters, rules of origin and local content requirements also now require wafer processing within the EC to qualify as local manufacturing. The net result will be increased investment by foreign multinationals in Europe. This definitely favors the Japanese who are much more flush with cash than any other foreign competitors.

France, with the help of Italy, has been fighting a lonely battle to slow penetration of the European market by the Japanese. Claiming that Europe cannot afford what has happened in the U.S., French Prime Minister Edith Cresson has been waging the battle primarily in autos. Currently, quotas in France and Italy allow Japanese imports to account for no more than 3% and 1%, respectively, of the local markets. So far, this has sharply reduced Japanese market penetration compared to other European countries where no such quotas exist. Originally scheduled to disappear by the end of 1992, the French and Italians have been able to negotiate a six year phase-out period, something that is being fought by Japanese negotiators as well as Britain and Germany. Cresson has also been fighting to have automobiles assembled in Japanese auto transplant operations in Britain ruled as imports rather than as part of the EC under the rules of origin. (One can imagine how this sits with the British and Japanese!) As we said earlier, it will take time to finally determine just how open "open" the European Community will be.

Taking all of the above into consideration, one can speculate on who the long term winners and losers will be. Winners will certainly include well established U.S. corporations who are willing and able to match Japanese and local investment. This includes most of the U.S. computer firms led by IBM, and in autos General Motors and perhaps Ford. Less sure is how U.S. IC manufacturers will fare against the Japanese onslaught. Japanese corporations in autos and ICs will certainly benefit, primarily because they are able and willing to make the necessary long term investment.

A number of local communities and regions within the EC will also be winners, at least as far as the economic environment and jobs are concerned, as the result of heavy foreign investment.

Without massive restructuring and mergers of major corporations within the EC, EC 1992 may make life easier but will do little to improve the competitive situation in major industries such as autos, computers, and semiconductors. Competitive deficiencies in key electronic industries will also eventually negatively impact such currently "healthy" industries as telecommunications, machine tools, robotics, and consumer electronics. Even if "Fortress Europe" is the initial result of EC 1992, such barriers will eventually fall to the inevitability of global competition.

The Future

So what does Japan's future look like? Its most serious challenges appear to be internal, as it wrestles with problems of affluence (such as a weakening of the work ethic among younger Japanese) and an aging population. The emerging Asian industrial nations seem likely to take away certain mature Japanese industries, but these are industries the Japanese do not seem especially interested in keeping anyway. Moreover, much of the growth throughout the other Asian nations is heavily financed by the Japanese, so those profits will find their way back to Tokyo. The European Community seems ill-equipped to fight U.S. and Japanese competition in electronics, and apparently will rely on joint ventures and quotas as its main weapons. Taken as a whole, Japan seems to have the resources to meet its challenges and maintain world dominance in electronics.

However, nothing is certain. After all, who would have believed thirty years ago that Germany would be reunited and the Soviet Union would be just a memory? It may be that thirty years from now books will be written about how the Japanese lost the electronics industry.

CHAPTER 12

Toward the Future

We have examined a lot of issues in the preceding chapters. While such issues are important, there is nothing we can do about the past. In this chapter, we will examine what can be done in the future to revitalize the U.S. electronics industry.

The Semiconductor Industry

When I first started on this project, I wanted to limit my comments on the semiconductor industry. This was primarily because it has been analyzed by so many others, including a number of the authors already referenced in this book. This proved impossible, both because of my own background and interests and because semiconductors are the linchpin of all electronics. Erosion of the U.S. position is one of the most serious challenges facing this nation, although there are voices that say we have nothing to worry about. Our primary challenge comes from Japan and, therefore, this section seems an appropriate place to debate the different points of view.

Almost everyone has an idea about what should be done about the U.S. semiconductor industry. On one side are those who believe that production doesn't matter because of a coming worldwide glut in wafer fab capacity. According to these people, only the skill of our designers, application engineers, marketing personnel, and sales people is vital. On the other side are many more who feel that the viability of the U.S. electronics industry depends on a vital merchant IC industry which controls its manufacturing as well. This debate is really about capital—that is, the relative importance of human intellectual capital compared with the increasingly massive investments required in capital equipment for the next generation wafer manufacturing capability. Such authors as Gilder in *Microcosm* and, more recently, Rappaport and Halevi in "The Computerless Computer Company," an article in the

July-August, 1991 issue of the *Harvard Business Review*, argue that intellectual superiority, not manufacturing expertise, is what counts. Conventional wisdom, along with most of the leaders of the semiconductor and computer industries and study groups such as the MIT Commission on Industrial Productivity, argue the contrary. There is no question which way the Japanese are going. They wish to control the entire process: the starting raw materials and production equipment, device and systems design as well as manufacturing, and worldwide distribution to the customer. So far, this approach is winning.

The intellectual capital argument is compelling since it makes best use of current U.S. strengths, is relatively "cheap," and allows U.S. companies to maintain their lead in the end-product industries as they are losing it at the front-end (that is, manufacturing). Rappaport and Halevi go even further. In addition to a "fabless IC industry," they argue for a "computerless computer society." By this, they mean that the computer industry should subcontract its hardware manufacturing to outside, generally offshore, vendors and add value through creative software development and user applications. It seems that many want to get us out of the manufacturing business.

In my opinion, there are fatal flaws in such an argument. Who can say that our intellectual lead is maintainable, especially with our basic education problems, the mobility with people within U.S. industry, and the large number of foreign students that attend our engineering graduate schools? It certainly hasn't been in the semiconductor industry, which belonged solely to the U.S. at one time. It is intellectually inconsistent that a nation which still maintains the greatest research capability in the world must be relegated to letting other nations capitalize on much of it. (Following this argument, biotechnology will be the next to go.) Also, the manufacturing floor has proven to be a major source of innovation. This source is largely cut-off in the absence of a manufacturing capability. Another objection is that there *is* a relationship between hardware, software, and IC design. Companies that combine all these capabilities will have the advantage. There is also a relationship between IC processing and the performance of the resulting devices. One quickly learns this while trying to match the design rules of a contract wafer fab with the desired performance of the circuit.

If contract wafer fab capacity is to be available so cheaply, who will be in a position to fund the next generation investment, whether they are from Taiwan, Korea, Singapore, Hong Kong, China, or wherever? The drive for ever smaller feature size will continue; those that control this will ultimately win the game. This is why large, keiretsu-based Japanese industrial groups plan to do the whole job. Perhaps most importantly, in no way will such a strategy lead to replacement of all the jobs lost in the U.S. semiconductor, computer, and related industries.

I cast my vote against the arguments of people like Gilder, Rappaport, and Halevi. The U.S. semiconductor industry cannot survive without a strong manufacturing base.

What exactly is the "semiconductor industry" anyway? Table 12-1 gives a quick "portrait" of it. The dozen companies at the bottom of the table are all "merchant" suppliers, meaning they sell semiconductors in the open market. The Japanese dominate these rankings, in no small part due to the financial strength of their parent companies and the monetary and other relationships that exist within the associated industrial groups.

From Table 12-1 we can conclude that there are roughly three basic groupings of semiconductor companies, although only two are shown. The first is companies that make semiconductors on primarily a "captive" basis and use them to enhance their competitive position in the end products they supply. By far the most powerful of these is IBM which—despite financial problems, restructurings, and slow growth—

TABLE 12-1

Comparison of Major Semiconductor Companies

Merchant Semiconductor Companies

Company	*Rank*	*Total 1991 Semiconductor Revenues*
NEC	1	$4,830,000,000
Toshiba	2	$4,680,000,000
Hitachi	3	$3,780,000,000
Motorola	4	$3,200,000,000
Texas Instruments	5	$2,900,000,000
Mitsubishi	6	$2,590,000,000
Fujitsu	7	$2,490,000,000
Intel	8	$2,400,000,000
Matsushita	9	$2,050,000,000
Philips/Signetics	10	$1,680,000,000

Captive Semiconductor Companies

Company	*Total 1990 Revenues From All Sources*	*Profits*
IBM	$69,020,000,000	$6,020,000,000
AT&T	$37,290,000,000	$2,736,000,000
DEC	$13,910,000,000•	($871,000,000)•
Hewlett-Packard	$13,230,000,000	$739,000,000
Hughes Electroncs	$11,730,000,00	$726,000,000

• 1991 figures

should be able to challenge anyone in the semiconductor world. In addition, IBM will continue to be one of the largest consumers of ICs purchased in the open market. I believe that IBM should also be able to serve as a source of investment funds, not just for internal growth but also for entrepreneurial start-ups. They could also provide foundry services on a selective basis, as AT&T has done for Western Digital. While less strong than IBM in semiconductors, AT&T should be able to compete globally because of their overall financial strength. At least based on their numbers, both Hewlett Packard and General Motor's Hughes Electronics should be in good shape, although the dire straits of the U.S. auto industry must cause concern for Hughes.

On the other hand, DEC may find it difficult to continue to fund what must be a very expensive captive IC capability. This effort first started during the 1970s in the Sprague Worcester facility when I was plant manager. They occupied space and leased equipment that became available when Mostek moved its wafer fab to Carrolton, Texas. At the time, we were all amazed at how inefficiently DEC ran the facility with no apparent consideration for cost (perhaps protection of proprietary information justified the expense). However, in my opinion, a number of OEMs such as DEC would have been better served, as would have been the entire IC industry, if they had chosen to create special relationships with merchant suppliers rather than trying to set up their own capability. (I must admit I was highly unsuccessful in selling this concept at a time when the fortunes of the end equipment companies were running so high. It might be easier today!)

Where the real competitive battle is being fought is within the second grouping, namely the large merchant semiconductor houses. In this country, these include Motorola, Texas Instruments, Intel, National Semiconductor, AMD and several others. Harris, Cypress, and LSI Logic also belong in this group. Intel seems well positioned due to its proprietary product line and penetration of the Japanese market. This is also true of Motorola, because of its strong semiconductor position and its world leadership position in mobile communications, especially cellular telephones. However, based on financial performance, most of the others are looking at a rocky road. *Made in America* by the MIT Commission and *Technological Competition in Global Industries* by David Methe both conclude that only large firms can compete in the global IC business. This is where the Japanese advantage lies, especially with the large computer/IC firms that are part of a keiretsu. I agree.

Answers? Restructuring, mergers, consolidations, use of foundries, and cooperative efforts will all be tried but only a few will succeed. One possible alternative to merger or consolidation would be for a group of firms to jointly fund and use the next generation wafer fab, sort of a controlled foundry approach. As with all foundries, there is a real problem with matching the fab processes and the performance needs of the different product strategies. Another difficulty relates to maintenance of proprietary information. Perhaps most important are the problems that would arise because of the different cultures involved. Just like Japan, getting fierce competitors to work together can be next to impossible.

As suggested earlier, another possibility is equity investment by major users, preferably, but not necessarily, limited to U.S. corporations. The investment in Intel in the 1980s by IBM is a good example, although how beneficial the results were is less clear. A variation of this is outright buy-out by large industrial firms that are either looking for diversification or who feel that IC capability is critical to their future growth and success. To say that I am negative on this solution is an understatement. Michael Porter's "From Competitive Advantage to Corporate Strategy" published in the May-June, 1987 issue of *Harvard Business Review* documents the incredibly poor results of diversification by acquisition on a non-industry specific basis. Closer to home, the results of the acquisition of Fairchild by Schlumberger and Mostek by United Technologies can only be classified as disasters. In both cases, the buyer had no understanding of the business they were acquiring and the different cultures involved.

"Industrial teams," as the relationships between IBM and Intel as well as Apple and Motorola have been called, really seem more like preferred vendor programs than the clustering that occurs in Japan. As previously discussed, the continual erosion of the U.S. semiconductor infrastructure also poses a major problem as far as creation of true industrial teams is concerned.

One thing is for certain: one way or another, a massive restructuring and consolidation of the U.S. semiconductor industry is required—and soon. Without this, a number of the companies involved will downsize, be bought, or fail. The National Advisory Committee on Semiconductors (NACS), in its initial "Micro Tech 2000" report, has predicted that by the year 2000 just ten companies of at least $2,000,000,000 each will supply all of the world's demand for advanced ICs. Less than half will be in the United States. I would not go that far, because there is a wild card in the deck: the third group of companies, which range from new start-ups to firms or divisions with semiconductor revenues up to several hundreds of millions of dollars. It is within this group that many of the new innovative products and technologies incubate. (Now I sound like George Gilder!)

By definition, these companies are specialists. They may concentrate on design, modifying their design rules to match available foundry capacity. Or, if they have special processing capability (as required, for example, in Smart Power and other merged technologies), they pursue product strategies where feature size is not the dominant factor (they just can't afford it). For some reason, many of the names start with V (Vitelic, Vitesse, VLSI Technology, VTC Inc.), X (Xicor, Xilinx, XTAR), or Z (Zilog, Zoran, Zymos). There are also some familiar names which have been around a while, such as Analog Devices, Burr Brown, Cherry Semiconductor, IRC, Raytheon, and our old (new) friend the Sprague Semiconductor Group (Allegro Microsystems). By the time this goes to press, some firms in this third group will be gone, new ones will have appeared, and there will be some new millionaires.

It would be an interesting exercise to study the different strategies followed within this last group. Most at the embryonic stage are thinking only of growth. Others have elected more gradual expansion, choosing to always remain specialists in products and markets where they can maintain a differentiated position and competition is limited. Still others have opted to phase out lines as they become commodities and pursue the next new niche. To a large degree, this was the Sprague approach since the late 1960s. While Sprague was never a major factor in semiconductors, $100,000,000+ in revenues is nothing to be sneezed at.

A start-up with a good team, a differentiated product offering, and reasonable financing is almost guaranteed early success. Much like the transition from adolescence to adulthood, the move from a group 3 to a group 2 company is more difficult. Staying there is even harder. As the business matures, competition becomes more intense, financing more demanding, and staying on top ever harder. The key to success moves away from the innovation stage toward the manufacturing floor. In this context, the skill of the Japanese in process engineering, manufacturing, and quality have repeatedly proven to be a competitive advantage that is hard to overcome. Add to this a largely closed Japanese market, the keiretsu system, and a lower cost of capital, and the mature end of the semiconductor market becomes an increasingly difficult one in which to succeed. With rare exceptions like Intel and Motorola (with its added strength in mobile communications) few others have succeeded in recent years.

Consolidation and restructuring are one answer. Why not let the Japanese buy more American firms? As far as the U.S. economy and jobs are concerned, this is not necessarily bad. As with many other industries, Japan and the United States need each other in semiconductors. The U.S. and Japanese economies are the world's two largest, and both nations need access to the other's markets. Despite their increasing move toward independence in technology, the Japanese still need to tap the uniquely innovative capabilities of U.S. start-ups. And, more than the Japanese need our ICs, the U.S. electronic original equipment manufacturers need theirs. What would make most sense as consolidation continues is joint equity arrangements and joint ventures between U.S. and Japanese companies. This would also help give us better access to their markets. But, as already pointed out, outside investment in the U.S. is easy, while in Japan this has proven next to impossible. Without some change in this situation, I see no answer but a gradual closing of U.S. markets to the Japanese. I do not believe either country can afford to let that happen.

What role can the U.S. government play in this? Assuming that some industry consolidation must occur, our anti-trust legislation must allow this, despite the uproar which may come from individual firms which feel that this is not in their individual best interests. Far better that U.S. companies consolidate than only be purchasable by foreign corporations, or fail.

Pressure must continue to further open up the Japanese market and, once open, U.S. suppliers must practice the necessary initiatives required to serve this market. Much has been written about the 1991 semiconductor trade agreement, most of it negative. The goal of 20% of the Japanese market is just that—a goal. There are no teeth to really make it happen, and there are even questions on how to measure it. For the moment, the U.S. original equipment manufacturers are probably satisfied because they will continue to receive the Japanese DRAMs and other parts they need. This comfort may wane, however, as threats of price hikes are raised. If, in a few more years, the remaining U.S. memory houses are gone, discomfort will undoubtedly turn to panic.

In *Technological Competition in Global Industries*, Methe has evaluated the importance of various research cooperatives, including the Pentagon-sponsored Monolithic Microwave and Millimeter Wave Integrated Circuit (MIMIC) and Very High Speed Integrated Circuit (VHSIC) Projects, and the more industry-oriented Microelectronics and Computer Consortium (MCC), Semiconductor Research Cooperative (SRC), and Sematech. The defense-oriented projects appear to be focused more toward systems and system interfaces than components, and therefore of questionable impact in the commercial IC market place. Sprague was a member of MCC for a while, and most of the reports were Greek (at least to me).

As far as SRC and Sematech are concerned, my only direct experience is with latter, which seems to have been surrounded by controversy from the beginning. The latest brouhaha at the time this book was being written concerned a possible refocusing toward CIM and software. This would mean elimination of the original "Phase 3," which was for development of a 0.35 micron manufacturing process, in favor of greater emphasis on modeling and software. The pressure for this change comes from large Sematech members such as IBM, AT&T, Texas Instruments, Motorola, and Intel who already have submicron capability.

Other than the financial and human resources committed to these cooperatives, there seems little harm done and certainly some basic knowledge and capability have been developed. However, one must question their true value. Transferring technology within a corporation is hard enough. Trying to do so from a research cooperative seems close to impossible. The corporate funds allocated to them probably would be better spent internally where, if anything useful is to be accomplished, they must eventually be spent anyway.

There is one other solution to the U.S. semiconductor problem that no one seems to want to discuss. It involves U.S. equipment manufacturers that buy semiconductor devices. Since the late 1970s, they have turned increasingly to Japan. We push for a "buy American" initiatives in Japan, so why not one in the U.S.? (Think what this would also do for the embattled auto industry.) If the answers are that U.S. companies just can't compete, don't have the capacity, or don't have the right types of products, then the game is over anyway. If we can't serve the U.S. market, how can we expect to meet the even more demanding Japanese requirements? No one wants a free ride but, to a large degree, the future lies in the hands of the users. Only they can create the industrial teams required long term to survive both at the component and at the equipment levels. It certainly seems to me that this solution is worth an "all court press." That is, unless no one really cares.

Revitalizing the U.S. Electronics Industry: The Macro Perspective

I believe that some things necessary for revitalization of the U.S. electronics industry require action by the federal government. If I had a "wish list" to give to the President and Congress, it would look like this:

- Reduce the federal deficit as one of the principal ways of lowering the cost of capital.
- Create a spirit of cooperation between government and business; the administration in power must understand the importance of a vital domestic economy.

- Embrace the philosophy of free trade while working aggressively to equalize the playing field with foreign competitors. This includes penalizing violations such as dumping, protectionism, and flagrant breaches of trade agreements as well as working toward a minimum share of 20% of the Japanese semiconductor market by U.S. firms.
- Provide federal support for industrial research and development at both the university and corporate levels with an increased emphasis on applied research and commercialization of technology.
- Increase federal support to improve education at the pre-school through secondary levels, with a special emphasis on mathematics, statistics, and the sciences.
- Improve protection of U.S. intellectual property by accelerating the related review systems in the courts.
- Stimulate the long term investment and retention of equities by eliminating or reducing the capital gains tax.
- Modify anti-trust legislation to permit intra-industry consolidation within, for example, the U.S. components industry. Require a more careful review to make sure foreign investments in the U.S. do not harm U.S. national interests in both military and civilian industrial sectors.
- Make sure that federal standards in such areas as waste disposal, pollution, safety, and gas mileage are enforced fairly, expeditiously, and within realistic financial limits. The same can be said relative to product liability.
- Introduce an effective national health care system as a means of protecting all citizens while controlling spiraling health costs.
- Place local content requirements on foreign manufacturing subsidiaries in the United States.

Revitalizing the U.S. Electronics Industry: The Micro Perspective

Executives and managers of U.S. electronics companies might agree with my "wish list," but they can't sit around and wait for the federal government to come riding to the rescue of their companies. However, there are several things they can do to help their companies meet the Japanese challenge. Not surprisingly, most of these things are already being done by their Japanese competitors. If I were CEO of an American electronics company today, here are the things I would be doing:

- Create and maintain the most competent senior management team possible.
- Depending on the size and maturity of the business, have one or more enabling and sustainable technological advantages over the competition that are maintained at all costs. Avoid unnecessary and unrelated diversification.
- Run a bare-bones operation that makes use of every available opportunity and sets sustainable long term profitability as a key corporate goal. Tie all compensation directly to total corporate performance and have profit sharing cover every employee.
- Utilize cross-functional teams for product design and development that combine the best of U.S. innovation and Japanese manufacturing technology skills. Make key suppliers partners in this activity.
- Pursue a major customer strategy with a global orientation. If a new or emerging company, pursue a joint venture strategy overseas.
- Making total quality control an ingrained element of the entire corporate entity, and I (the CEO) would be its champion. Continuous incremental improvement (kaizen) would be the cornerstone of this effort..
- Employ every possible means to encourage worker inputs and involvement, including practicing cross-functional training throughout the corporation.

CHAPTER 13

NEWCO

I suppose we would all do some things differently if we had them to do over. I'm no exception. Experience is a valuable teacher, and everything is obvious in hindsight.

Just how does a company approach a business strategy that can survive in today's environment? There is little to be gained by restating the many lessons already covered in this book. Instead, I would rather take the approach of rewriting some history and creating and going through the early growth phases of a hypothetical company based on many of the principles already covered. To make things more interesting, fact and fiction will be intermixed along with the names of real and imaginary (in quotation marks) participants. To the former, I express my apologies for changing their lives. Hopefully, these changes are only beneficial. In no way do I claim that NEWCO is the right or only way. It is, however, based on taking a bare bones approach and on taking advantage of every unique situation available. This is the way many successful start-ups proceed and succeed.

At any rate, if I had it to do over, here is how I would do it. . . .

As the economic environment in electronics ground listlessly through 1986 and Sprague Electric continued its massive restructuring, it became evident that my days as president and CEO were numbered. Penn Central was increasingly disillusioned with the electronic components business and wanted their own man at the helm. In addition, key growth areas such as semiconductors and ceramic capacitors were heavy financial drains with little short term possibility of improvement. Private discussions I held with several different investment banking firms all gave the same answer: the operating cash flows were insufficient to support any form of leveraged buy-out from Penn Central that made sense from a business standpoint. Nor could I find a way of spinning off the semiconductor group, my real first love, for the very same reason. A report from the Boston office of McKinsey, contracted by Penn Central, also recommended that Sprague exit the MLC business as expeditiously as possible, even through this represented the sole growth segment in the core capacitor business. Even if I could somehow survive, running Sprague Electric had long ago lost its allure. What I really wanted to do was to start a new business—that is, if I could come up with a reasonable product concept.

I figured that I had certain things going for me. Since it was clear that Penn Central wanted a smooth and non-confrontational way of changing management, they would probably be inclined to award me my full severance package. This included two years salary plus bonus which, before taxes, would yield about $300,000 each of the first two years after I retired. There was also a one year consulting agreement with a fee of $100,000. I intended to personally invest at least half of the salary and bonus compensation in whatever new business I decided to pursue plus an additional $100,000 of my own funds, and I was sure I could raise another $100,000 from family and friends ("f&f"). Therefore, depending on how the severance was handled, without additional outside financing, an initial capitalization of at least $500,000 should be possible. There could be even more if I was willing to risk more of my own money. I intended to start absolutely bare bones and stretch these funds as far as possible. I also felt that Sprague Electric might be willing to trade certain technology and service elements I would need for some part of my severance package. This would have to be part of the negotiations. Because of my extensive business contacts throughout the electronics industry, I felt I could open a lot of doors not easily available to many start-ups. One thing I had learned at Sprague was clear: whatever the business or wherever it is practiced, personal contacts can be a critical element in success.

But what business strategy could be both successful and fun? I first considered doing something in integrated circuits, but quickly dismissed this as too capital intensive and competitive. In addition, I had to admit that I really had no unique ideas on product direction. On the other hand, Sprague was also disillusioned with ceramics, despite a leadership position in many ceramic formulations and the fact that MMCC had demonstrated that our build-up or "flip" process also offered some unique competitive advantages. Therefore, there might be some unusual opportunities in electronic ceramics if some of the Sprague capabilities could be effectively harnessed.

To check out this hypothesis, I invited two of Sprague's most competent scientists, consultant Sid Ross and research director Galeb

Maher, to meet with me for an afternoon of brain storming. During this meeting I said nothing about my intentions, although this certainly would be necessary if I actually went ahead with my plan. The question we wrestled with was how to develop the world's "best" MLC capacitor. Rather than try and cover all the different TCC (temperature coefficient of capacitance) ranges, we decided to concentrate on the X7R segment which allows a zero bias temperature variation of capacitance of no more than + or -15% over the full military range of -55 to + 125°C. To optimize both volumetric efficiency and dielectric breakdown, we needed ceramic dielectric layers that were simultaneously thin and near theoretical density. This meant using very high purity, chemically synthesized ceramic starting powders and a wet process similar to the Sprague "flip" system.

Sprague already had several proprietary formulations that met the TCC requirements and also had low voltage Ks (dielectric constants) as high as 3500. However, based on work reported in the U.K. and elsewhere, to achieve leadership dielectric constants close to 5000 would require both modification of the additives used in the $BaTiO_3$-based Sprague formulations as well as fine powders with a very tight sintered particle size distribution around 0.8 microns. As reported in the MMCC report, we believed that with proper control systems consistently depositing dielectric layers with a sintered thickness of 5 microns should be possible on a production basis. On the other hand, such thin layers would require the metal electrodes to be less than 1 micron and, although certainly feasible, doing this on a production basis would require more work.

The key technical question revolved around how to make the fine and uniform particle size $BaTiO_3$ powder. If this could be established, we were sure everything else could be accomplished through careful characterization of the related processes and good engineering.

Ross said that he felt the new Sprague emulsion process probably had the highest probability of success. An emulsion is a liquid, such as milk, where one immiscible liquid is dispersed in another. In the Sprague approach, the ceramic precursors are dissolved in an aqueous solution and emulsified in the appropriate organic fluid. Each of the individual droplets contains the chemical elements necessary to synthesize the final desired material which, in this case, would be $BaTiO_3$. This material is formed as a uniform fine powder when the emulsion goes through a series of heating steps. High purity starting materials insure the purity of the final ceramic. Sid said that careful control of the basic chemistry, organics, and heating steps should yield the desired powder size and tight distribution. "0.8 microns + or - 0.1 micron?" "I believe so." "How long to demonstrate feasibility?" "Three to six months of full time effort." "Want to try it?" "You bet!" Having defined the technical scope, they both asked when I wanted to start. I replied, "I'll let you know." They both wore perplexed looks when we parted.

Before approaching Pug Winokur and Al Martinelli at Penn Central, I attempted to determine what I would do if they would not agree to my full proposal and more up-front capital was required. After discussions with my son Bill at Kidder Peabody and several friends in the venture capital business, I determined that I would probably be able to raise the necessary up-front capital through the venture capital route, admittedly at a much higher cost than my own self-funding approach. Either way, I would first have to demonstrate product feasibility. Fund managers were also nervous about the fact that it appeared I was planning to compete head-to-head with giants in this country and in Japan such as AVX and Murata. However, I was able to convince them that I had some ideas up my sleeve here as well. However the funds were raised, the key was obviously access to the Sprague technology and its extensions. Without this, I would have to find a different product thrust. So off I went to the Penn Central Headquarters at 245 Park Avenue in New York City to meet my new destiny.

My proposal was relatively straight forward. In exchange for taking early retirement and giving up one year of my severance compensation, I wanted the following: the one year $100,000 consulting agreement as originally provided in my severance package, a royalty-free, non-exclusive license under the Sprague ceramic patents and related know-how for ten years, one year free rental of a single bay in the Sprague Research Center plus free use of a full-time technician for the same period, and free use of library and analytical services for two years for a total expense not to exceed $50,000. While this might seem an unequal trade for Sprague, it actually saved them money compared to the complete severance package. The exchange of licenses and services for the one year severance cost Sprague nothing on an out-of-pocket basis. In addition, unknown to the Penn Central executives, I had also arranged a back-up alternative to the space and technical support requests in the Williams College chemistry laboratory. This was possible since I still owned a house in Williamstown, was a long time financial supporter of the institution, and was also the father of one graduate and a current sophomore. As long as there was no conflict, I also proposed that I be allowed to employ Ross as a consultant—that is, if he agreed. If Galeb should balk at my proposal, all deals were off and I would go into the consulting business under my full severance arrangement. I didn't add the obvious: without the Sprague ceramic technology I also would have to abandon my plan.

Much to my surprise, Pug and Al not only agreed to most of my proposition, they actually seemed enthusiastic. Oh, we had to dicker since this is always part of the game. In a nice gesture, they asked me to remain on the Sprague board. However, my consulting was reduced to $50,000 in return for which I was able to extend the free space rental and technician use for a second year. I agreed to provide Sprague nonexclusive royalty-free licenses under any NEWCO patents for the same ten year period and agreed not to compete with Sprague in any product areas outside of ceramics for two years. I was also able to purchase an excess "flip" machine from the Sprague Wichita Falls plant for the rock-bottom price of $25,000. As far as working in ceramics was concerned, they showed surprisingly little concern about potential conflicts, apparently since they had already decided to exit the business. Perhaps they really hoped I might eventually purchase the entire Sprague ceramic business unit (something furthest from my mind). Even more likely, they probably felt my new enterprise would never get off the ground anyway. Galeb eagerly accepted the arrangement and said he would provide all the support he could. I was to learn later that he had his own ideas about the future of the labs when, in March of 1990, he bought the research facility and formed MRA Labs, thus becoming both my landlord and the owner of the patents I was licensing. Sid said he wanted to work full time on the emulsion process and, since "working was more fun than sitting around," offered to work the first year for the bargain price of $60,000. In addition, he wanted to take half as deferred compensation. I also promised he would be the first participant in our stock option program once it was set up. Therefore, on January 1, 1987, NEWCO was incorporated with one employee (me), Sid Ross as a consultant, and "Howie," our technician on loan. I planned no further additions until feasibility of the modified emulsion process was proven. In addition, to minimize the initial cash requirements, I planned to take no compensation for myself the first year, using the Sprague consulting fee as deferred compensation. Sid did the same, and we both recognized that actually receiving these funds depended on the success of the enterprise.

Because of the trade of licenses and services for one year of my severance compensation, the initial balance sheet was very unusual, as seen in the simplified version shown below. It does show the obvious: even with the licenses and "free" services, cash flow would have to be tightly monitored. The real key depended on how quickly feasibility could be demonstrated.

NEWCO Balance Sheet (1/1/87)

Assets:		
Cash and Equivalents:	$175,000	
Equipment:	$25,000	(5 year depreciation.)
Licenses:	$165,000	(10 year depreciation.)
Services:		
Space rent:	$25,000	(2 year depreciation.)
Technician:	$60,000	(2 year depreciation.)
Anal./Lib.:	$50,000	(2 year depreciation.)
Total Assets	$500,000	
Liabilities:	0	
Equity:		
50,000 shares common:	$500,000	(JLS 40,000 and "f&f" 10,000)
Liabilities and Equity:	$500,000	

While Sid and "Howie" were emulsifying, I began to plan for the future. While I knew that a business plan would be necessary shortly, I spent little time on this initially since we were still a long ways from product feasibility or requiring additional outside funds. I did need to begin recruiting the remainder of the management team which, in addition to Sid and I, would require, at a minimum, a head of marketing and sales and—especially important—a head of manufacturing. Because of the importance of process engineering and control, I was hoping I might be able to attract one of the experienced Japanese who were in this country in advanced degree programs. Contacts with some of my friends and associates at MIT, Penn State, Rutgers, and Lehigh identified several interesting possibilities. My contacts also tried to interest me in direct support of graduate research, although I said that would have to wait a few years. Finally, rather than a CFO (chief financial officer), I hoped to recruit an experienced controller, possibly from one of the Sprague business units or manufacturing locations. I decided early on that, at least at this stage, using a "headhunting" agency would probably not be required and I would first tap my personal industry contacts.

I also began to make some very preliminary customer contacts. As far as the latter were concerned, I felt that the best approach was to target a small group of potential key customers, including IBM, DEC, AT&T, H-P, Delco, and several corporations in the Route 128 technology belt. It also helped that I had excellent contacts at each. While any one of these companies would quickly overwhelm our limited resources, I felt that qualification would take time and, if successful, would insure our early success. What we had to sell was "the world's best MLC capacitor," that is, the unit with the highest volumetric efficiency, tightest X7R TCC, greatest breakdown capability in volts per micron, and, due to the density of the dielectric and total construction, highest potential reliability. In addition, while we would be qualifying a capacitor that best fit the definition of a commodity, I hoped that with customers such as these we might find one or more applications where our unique capabilities could solve higher margin special requirements. Just as the DRAM is the technology engine for the IC business, so would be our MLC. We were also following one of Sprague's early lessons: if you can succeed in serving the leaders, the other customers come much more easily. With a couple of notable exceptions, including Delco which continued its major supplier thrust, each of the contacted companies expressed polite, somewhat reserved interest, and agreed that if the claimed advantages could be demonstrated at competitive prices, they would institute immediate qualification testing. The problem was that we had no units, and probably wouldn't until later in the year. They said they could wait. If they hoped we would quietly go away, they were soon to be first disappointed, and then increasingly excited that we didn't.

Technical progress was steady, but agonizingly slow. Then, late in May Sid called and declared excitedly, "I think we've got it!" I rushed down to the lab where he and "Howie" showed me an innocuous beaker of powder which, at first glance, told me nothing. However, it was the charts of comparative particle size distribution that told the story: the NEWCO material was tightly clustered between 0.2 and 0.3 microns, much tighter than anything we had ever seen, either synthesized within Sprague or available in the open market. Proper sintering should yield the desired tight distribution peak around 0.8 microns. We went over and over the process and data to determine whether this was a fluke, or if reproducibility on a production basis should be possible. Repeated successful experiments indicated the latter.

With powder and data in hand, we went to see Galeb and asked if he could run some of the material in his pilot flip machine to try and make a 0.1 mfd size 1206 chip. With some refinement of the process and further reduction in dielectric thickness, we calculated that we should eventually be able to put 1 mfd in the same size, a real breakthrough. He said he still needed to do some further experimentation on electrode thickness control, already one of his current programs for the Sprague locations in Wichita Falls, Texas and Hudson, New Hampshire. He also asked that Sid and "Howie" continue to work on optimization of the characteristics of the ceramic paint he would use in the flip. Since the final results could also be very important to Sprague, he agreed to make the runs for free. All the usual delays occurred and then some. First, several lots bombed out, and then there was an equipment crash. Further work than expected was required on both the ceramic paint and the electrode paste, and the drying and sintering cycles had to be modified. Methodically, with close to kaizen-like patience each was solved. Finally, in early September the first fully successful run was completed. A lot of several thousand pieces came out with measured Ks just below 5000, an extremely flat TCC well within X7R, dissipation factors below 3.0%, and a flash voltage capability of 100 volts per micron. This was repeated on 5 more successive runs. In addition, using accelerated life test procedures developed by Sprague's George Shirn, reliability looked far better than anything commercially available. Admittedly, the starting ceramic powder and MLCs had been prepared with tender loving care by some of the industry's most competent scientists, engineers, and technicians. Still, based on the MMCC lessons, we saw no reason why the same results, if not better, wouldn't be seen on a full production process. At the end of September we held a champagne party and dinner for our tiny staff plus Galeb and his key managers. The next day, with samples, data, and hope in hand, I went on the road to sell our wares.

If I expected enthusiasm, I was bitterly disappointed. It seems that everyone was beginning to tighten up on costs, and several companies said they it would be months before their tests could be made—months I didn't have. However, the component engineering groups within CCP (corporate component procurement) at IBM and DEC agreed to go ahead, as well as a relatively new workstation company, "Electro Systems" ("E. S.") headquartered in Billerica, Massachusetts. This last contact had come through the Electronics Industry Association which NEWCO had joined early in the year as an associate member for the bargain price of $1000. After two weeks of travel, I returned to North Adams, exhausted and frankly depressed by the reception I had received. There seemed little indication that the world needed another MLC house, even one with "the world's best MLC capacitor."

And then we waited. Polite calls yielded little information, and October and November drifted by. Work continued in the lab on refining the processes for both powder preparation and MLC manufacturing and, with little else to contribute until we heard the results of the customer evaluations, I worked at preparing several technical papers with Sid and with Galeb, both for publication in various journals such as the Journal of the American Ceramic Society and for presentation at the 38th Electronic Components Conference (ECC) scheduled for May of 1988. These would serve not only to help establish us as factors in the field, they also might lead to contacts in Europe and especially Japan, which was increasingly the hotbed of activity in electronic ceramics. Patent applications were filed in the U.S., Japan, the U.K., France, and Germany on our materials and process modifications and improvements. We realized that the beginning of a patent portfolio could become a critical asset further down the line. I also began to prepare a detailed business plan, an absolute necessity once we began to require any appreciable capital expenditures.

Obtaining useful market information proved much more difficult than anticipated, primarily because there was no way to really estimate the future requirements for high purity fine ceramic powders, or for the NEWCAP®s they made possible. In addition, we were really dealing with two very different potential business directions, one related to materials and the second to components. An abbreviated summary of the key data is given below:

- World $BaTiO_3$ Powder Usage: approx. 20 M lbs/year
- U.S. $BaTiO_3$ Powder Usage: < 8 M lbs/year
- Known, non-captive fine powder competitors:
  Sakai Chemical: capacity = 130,000 lbs/year
  Rhone-Paulenc: capacity = 200,000 lbs/year
- MLC Caps.:

MLC Caps.:	*$ B Sales*	*# B Sales*	*ASPs*
World MLC Caps.	2.8	168	$0.017
U.S. MLC Caps.	0.52	12	0.044
World SMD Production.	1.245	83	0.015
U.S. SMD Production.	0.231	7.1	0.033

Several things were clear from this data. First, it gave no real information on the future market for high purity fine powder $BaTiO_3$, except that current usage was still very low. Lacking this, we assumed that by the early 1990s perhaps 10% of the available market might be available to fine powders, assuming a price of $10 to $12 per pound. This conclusion assumed limited in-house captive manufacturing by major MLC suppliers such as AVX and Murata. Second, there were some formidable potential powder competitors, regardless of the size of the market. Third, modification of our purchased flip to make it a full production equipment with an annual capacity of around 130,000,000 units (depending on mix) would cost between $100,000 and $300,000, depending on how much flexibility we wanted. Even then, it would only scratch the world surface-mount device MLC market. This did not take into consideration the cost of a production emulsion process which, for a capacity close to Sakai's, would be close to an additional $2,000,000. Therefore, everything pointed toward a niche strategy, at least until we were up and running and major capital could be raised. With "consulting" income during 1987 of $50,000, a cash payroll of $30,000 (Sid Ross), and tight control of everything else, even with a P(L) loss of close to $170,000, we still finished 1987 with our cash reserves basically unchanged. However, assuming we would be at some level of production of powder, product, or both by mid-1988 meant that we could soon be way undercapitalized.

Whatever happened, we would need powder. Unwilling to even consider capitalizing for full scale production, I asked Galeb if we could modify and reactivate the pilot emulsion equipment which was sitting idle in the nearby Windsor Mill. Originally designed to make powder for Sprague, it would require considerable modification. We agreed to pay for this, as well as all the necessary ceramic precursors and chemicals, supplies, and utilities. We then hired our second employee, an extremely bright young chemical engineer named "Joe" from Rensselaer Polytechnic Institute. Sid, "Joe," and "Howie" then set about activating the facility to produce up to 10 Kg/day of fine powder, tight distribution barium titanate. Then, in late-January of 1988, we got our first big breaks at the customer level.

First, an old acquaintance from IBM CCP, "John Battle," called and asked, "Are those MLCs you left with us for real?" I asked what he meant. "They are the best thing we have ever seen," he said. "While we would really like to see them in full production in one of our mid-range systems, there is a critical application for an airborne computer at IBM Owego. The performance we have measured on your units, stress capability, and the potential volumetric efficiency we have calculated may be the only solution to the extremely tight density requirements. I can't tell you much about the application except that it is a capacitor array." Trying to remain calm, I took the name and telephone number of the appropriate contact at Owego and called as soon as he hung up. They wanted a full scale presentation early the following week. Then, in what seemed an incredible coincidence, later that afternoon I received a similar call from the pugnacious chief engineer at "Electro Systems," "Bill O'Neil." This time it was to develop a substrate for a custom IC with one or more buried decoupling capacitors. Prototypes were required by year end, pilot production in 1989, and, depending on customer acceptance for their new workstation, full scale production would begin in 1990. As with IBM, there were NRE (non-recurring engineering) funds available. I put down the phone stunned, sat for just a moment, and then raced down the hall to assemble our tiny team. If we could respond to these two opportunities successfully, we would be truly launched. The specialty niche that was developing also meant that both our pilot powder and flip capabilities would last us a good deal longer than if we were pursuing a primarily commodity thrust.

The IBM negotiations were difficult, because they generally preferred to amortize NRE over the production volumes. However, they eventually agreed to fund $100,000 up-front upon signing the contract and another $150,000 when 100 prototypes that met full specification were delivered sometime in the fourth quarter of 1988. The "Electro Systems" agreement was similar, although NRE funding would be done in three equal parts of $75,000 each during the year. Total related pilot production requirements for 1989 for these two customers alone was forecast at close to $800,000, and would eventually come close to this number. Other similar opportunities began to appear, especially after the joint paper Sid and I presented at the spring ECC conference, "Practical and Theoretical Limits in MLC Capacitors." Some included NRE, although not nearly on the same scale as IBM or "E. S." Our problem quickly became picking and choosing those with the greatest chance of success and highest return. Some Japanese firms were so excited that we had great difficulty scheduling the visits they wished to make and controlling their disappointment when a conference room was all we would show them. I was also personally running out of time to try and cover the entire waterfront and was relieved and especially pleased when an industry veteran and former Sprague employee, "Jack," left his job with a U.S.-based Japanese subsidiary to join us as senior vice president for marketing and sales. While he took a major cut in direct compensation, as did we all, the potential return of the stock option program we had hammered out early in the year more than compensated. The shares were exercisable at the same price as the founders stock, that is $10 per share. And as "Jack" put it, "time is running out for one more shot at the brass ring." The team was shaping up.

Before actively pursuing new financing, we had to decide whether or not we wanted to produce both powder and components. We calculated that the prototype emulsion facility available from Sprague would carry us at least through mid-1990, but after that we would need additional capacity, and that would be expensive. In addition, I was not convinced that long term the only way to make the type of powder we required was via the emulsion approach. It was the combination of the powder, the other materials and processes we used, and product design that produced the unique characteristics of NEWCAPs and their derivatives. Therefore, as far as competitive advantage was concerned, what we really had was lead time and whatever protection we could get from the patents that would eventually issue from our applications. After lively debate we finally agreed to license the powder technology, assuming we could find someone who could duplicate our work, was capable of raising the necessary capital, and would give us at least two years of exclusivity. Almost by definition this eliminated Japanese material suppliers since we were uncomfortable how long our results would stay proprietary. We also had to find someone who was acceptable to Sprague. The solution would come when Sprague placed all its ceramic operations as discontinued in mid-1989 and planning began by Galeb and his team to purchase the Sprague research facility and create a contract research laboratory. MRA Labs would be incorporated in March of 1990, with lots of excess space for rent, broad capabilities in materials, analytical and library services, and ownership of the Sprague Electric ceramic patents that we were licensing. MRA would also eventually become our powder source, although the financing of that is another story.

In addition, as the result of Sprague's pull-back in ceramics, the end of 1988 we were able to purchase a second Wichita Falls flip for 10¢ on the dollar. For the time being, we planned to mothball this equipment until it was required for added capacity. Similar flips were to become available even more cheaply when, in 1990, Sprague finally closed its Wichita Falls, Texas, MLC manufacturing facility. However, due to our increasing product thrust into customer specials, as with IBM and "Electro Systems," what we had to do first was to modify Flip 1 to be a flexible manufacturing equipment. To do this required hiring another Sprague alumnus, "Phil," who had experience in both MLC ceramic processing and equipment design. While the actual equipment work would be subcontracted, "Phil" immediately organized a cooperative design team involving, in addition to himself, Sid, "Joe," "Howie," a new master technician, and consultation from Galeb and his group. We had no intention of duplicating the mistake we had seen in all too many Sprague developments of designing products and equipment that were incompatible with the materials and processes required to make them. This approach of cross-functional development teams was to become a way of life throughout the company's growth.

Therefore, as we entered 1989, we had two chip manufacturing machines, the original one which we were modifying, and the second machine we had purchased and which would be held in reserve until needed. The NEWCAP capability brought competitive advantage in almost any application where volumetric efficiency and reliability were key. If this capability ever hit the commodity market, then major additional capacity would definitely be required. On the other hand, our early niche approach did not need this. Rather, in addition to performance, what we also had to sell was quick turn-around to customer requirements and flexibility. As long as this approach dominated our business, most our capital requirements were to modify and upgrade our two flips, chip handling, and especially test equipment.

The final key members of the management team were identified toward the end of 1988. First to join was "Mike," the controller from Wichita Falls who, sensing what was about to happen, left the ship before it sank. Originally an engineer in the labs in North Adams, he brought simultaneously a good understanding of technology, hands-on operational experience in ceramic components, and a keen analytical mind. Then, just before Christmas, I received an unexpected telephone call from a MIT graduate student, "Jimmie Yamoto," who had been referred to me by Kent Bowen. With many years experience at TDK in electronic ceramic manufacturing, he had been sent to the U.S. to broaden his skills in materials science and earn his MS degree. In the process, he had fallen in love with a fellow graduate student, "Marie." They were to be married immediately after graduation in June of 1989 and, being a Massachusetts native, "Marie" had no intention of returning to Japan. Love had apparently overcome culture and responsibility, so "Jimmie" planned to stay in the U.S. and wanted a job right after graduation. Kent said he was the best masters degree graduate student he had ever had. He was also to prove to be the most competent, disciplined, and demanding manufacturing head I had ever seen. After he joined NEWCO, all the well-publicized Japanese skills in total quality control, manufacturing engineering, and JIT were to become a way of life. Not that we didn't already know they were important. As a matter of fact, one of the key features of NEWCAPs was the superior reliability inherent in the materials, processes, and designs we used. On the other hand, "Jimmie" brought an extreme discipline and drilled into us the concept that TQC was a total corporate activity that started at the top. As a result, we never did hire a QC manager and I became the chief TQC facilitator. He also introduced a kaizen approach to problem solving that was unique to my experience. If Sid and I had provided the original direction and innovation, ultimately "Jimmie" made it all gel. Since "Marie" was an extremely competent materials scientist and wanted to work, we also hired her as well when they both became available. After MRA was formed, although she remained on our payroll, she went to work in their labs on development of improved formulations and on a detailed analysis of the different characteristics that resulted from different $BaTiO_3$ sources. These included both our own emulsion powder and commercial fine powder sources. Happily, so far the emulsion material had continued to demonstrate certain unique advantages.

But I am getting ahead of myself, and we need to return to the latter part of 1988. Successful prototypes have been delivered to both IBM and "Electro Systems," new opportunities are arising almost daily, "Jack" wants to start sampling NEWCAP MLCs on a very selective basis, and we are going to run out of funds sometime during late 1989 or 1990. To get a better idea what we might need, I ran a rough P(L) for 1988 through 1992 so that I could determine cash requirements. Since the year was nearly over, 1988 was pretty well fixed. I also was reasonably confident about 1989. However, after that I knew things were increasingly pure guesswork. For simplicity, in calculating cash flow my figures didn't consider receivables, payables, or inventories. (That is a real JIT system!) My estimates for 1991 and 1992 were admittedly arbitrary and, if anything, too pessimistic. The most important result was that, if everything went as planned and we were careful, the real cash bind wouldn't come until 1990. Still, it would be very tight.

With the help of a friend with a small desk-top publishing business, I put together a smart-looking plan with lots of market data, colored charts, projections, sensitivity analyses, and the like. There were pictures of some of the new products, an impressive comparison between the emulsion powder and commercially available material, and even I was impressed with the results. Admittedly, besides balance sheet information and much more detail on cash flow generation and requirements, in addition to what might be best described as boilerplate, there was not a great deal more than the basic information developed in my cash flow statement. Still, I felt the story ought to be impressive to any knowledgeable investor, particularly if he took the time to follow-up on some of the letters from current customers which were included. In a way, it was much like some of the detailed proposals I had evaluated while a several time review panelist on the National Science Foundation Engineering Research Centers (ERC) and state Industry University Cooperative Research Center (IUCRC) initiatives. One thing these reviews had taught me is that the impression the presenter makes is much more important than the details of the presentation itself. This is one thing that makes getting venture capital so difficult. Fund managers read literally hundreds of proposals before choosing a very limited few to review in person. As a result, unless you have contacts in the field you may never get a chance to show your wares in person. Because of this, before broad presentation I felt I should run the NEWCO plan by some people I knew well enough to provide an honest appraisal.

My son Bill, who was now at Smith Barney in mergers and acquisitions, was somewhat noncommittal. Although he had some excellent suggestions concerning format and information he felt should be either deleted, added, or changed, it was pretty obvious that he was not very excited by the whole idea. Ed Getchell, president of Venture Founders, a specialty venture capital firm in Lexington, Massachusetts, was more direct. While he was impressed by what we had already accomplished and were planning on doing, he stated flatly that he did not believe this type of business was an especially good candidate for venture capital money. Later discussions with other venture capital firms proved he was correct. Ed said that while venture money could be found, such investors want a quicker return on their money and that those who would commit would want too much of the company for too little investment. He added, "While your technical description is fascinating, it is too hard to understand what you have that is unique and, unlike the semiconductor business, it just isn't sexy enough." He suggested that we might like to look into some of the industrial revenue bond (IRB) opportunities available and stated that a company similar to NEWCO had good luck obtaining local support in Woonsocket, Rhode Island. Since so many of our resources were already in place in North Adams, I didn't see how we were in any position to look outside of western Massachusetts, at least for the near future. However, this was an interesting alternative to consider somewhere down the line. He then suggested that we look for a corporate partner, with current customers high on the priority list. IBM was too big, and I wasn't ready to deal with several Japanese companies that had already expressed just such an interest. Therefore, "Electro Systems" seemed like the most likely candidate. I made arrangements to meet with "Bill O'Neil" the following week. Somewhat surprisingly, he had been thinking along the same lines. As we were to learn later, they had determined that our capabilities were providing them some very unique advantages over their competition.

The negotiations were difficult and lengthy, and it wasn't until January of 1989 that a deal was finally struck. By this time, we were already in pilot production of their substrates, they were trying to accelerate the program, and several new projects were underway. They also wanted to design a memory board around our volumetrically superior MLCs. After hammering out the difficult exclusivity issues, contingency alternatives if we should fail to deliver, the revised make-up of the Board to include "Bill" as the "E. S." representative, a right of first refusal if they decided to sell their equity, and a host of other items, we signed on Thursday, January 26, 1989. We negotiated a value for NEWCO of $4,000,000 and sold "E. S." 25% of the company. This meant issuing them 16,667 shares of common for $1,000,000 and revaluation of the share value from $10 to $60 per share. As shown below, the founders had made a tidy gain at least on paper (for purpose of this example no consideration is given to dilution resulting from stock options).

Stockholder	*# of Shares*	*Initial Value*	*New Value*
■ J. L. S.	40,000 (60%)	$400,000	$2,400,000
■ f & f	10,000 (15%)	$100,000	$600,000
■ E. S.	16,670 (25%)	n.a.	$1,000,000
	66,670 (100%)	$500,000	$4,000,000

Up to the "Electro Systems" investment, the board of directors had been all inside people and the meetings pretty informal. This all changed once "Bill" was on the board. He was precise, demanding, and set on never harming the interests of "E. S." This made creation of new stock options extremely difficult and, at first, he completely balked at an employee stock option plan we wanted to introduce. He also had difficulty understanding why we were planning to spend so much time, effort, and money on in-house training and the creation of cooperative relationships with local educational institutions. We felt these were critical to our success once we were in the production mode, our staff warranted it, and the company could afford the cost. "It will be too damn expensive," he exclaimed in frustration at one of our meetings. However, as he became more knowledgeable in our business and began to understand how much importance we placed in our people, he began to acquiesce. Happily, in the end he became one of our strongest supporters.

A more difficult problem arose with our current and potential customers who were competitors with "Electro Systems." IBM was one, so was DEC, and several others. I personally visited every one and, in all but one case involving a possible new customer, was able to convince them that the ownership by "Electro Systems" would in no way compromise our ability to serve them effectively. As a matter of fact, it only strengthened our capabilities. In selling this concept, we had two things going for us. First, from the beginning we had assigned major account coverage to every company officer and key employee. As we grew we continued to follow this approach religiously. Therefore, we had excellent rapport with all of our major accounts and they had grown to trust us. Secondly, our capability was still so unique that they needed us and we worked very hard not to take unwarranted advantage of this edge. The TQC concept that we had started early on, and which was later refined by "Jimmie Yamoto," had customer satisfaction as a major component. As the result, meeting commitments to both internal and external customers was another company cornerstone. We knew from painful past experience that leadership technology, even when accompanied by superior quality, was no longer a guarantee for success. We also began to recognize that we were approaching the time when the lack of a viable second source in our critical specialty niche would eventually cost us business. Since both the marketplace and competition were global in nature, even at this early stage in our development we also realized that we could not for long neglect Europe and Asia. Perhaps we could kill two birds with one stone.

"Jimmie" and "Marie" joined us in July and immediately immersed themselves in gaining a detailed understanding of our materials, processes, and the equipment that used them. They had only limited suggestions concerning the emulsion process, but our flip system was another story. "Jimmie" was fascinated by the approach and had immediate ideas concerning how to make the curtain and screener systems more flexible. Fast changeover was becoming increasingly critical as our specialty business increased and diversified. He was also quick to point out that we needed to do more on cleanliness in the equipment space and drop-ceilings and laminar flow hoods were installed. His most important contribution, however, related to the driers. As it was, these were the rate-determining parts of the closed-loop system and, because of this, we were not making most efficient use of the total equipment. A continuous system might be "sexy" and appropriate when we were making MLCs of a few part types. However, when many different types of special devices were required, it was just plain impractical. As a result, when we began our modification of the second machine we had purchased, we unbundled the driers and bought several more which we ran in parallel with the rest of the equipment. While this meant we would need more operators to handle the plate transfers, the added equipment capacity more than compensated for this. Actually, "operators" was a misnomer. No one in NEWCO punched a time clock and everyone was, in effect, on salary. Those that operated the production equipment were completely cross-trained in the different operations and were much closer to master technicians than what we would have conventionally referred to as direct labor. The resulting improved productivity and yields more than made up for the added expense. Eventually, flip 1 would be used primarily for MLCs and, because of its even greater flexibility, flip 2 dedicated to specials.

By fall, the need for a second source was becoming critical as was the need to look at the overseas markets. With all the interest we had already generated, plus the excellent contacts both "Jack" and "Jimmie" had, a Japanese association seemed to make the most sense. Discussions with our major customers indicated that they had little concern about a Japanese second source since they were already purchasing many of their component requirements there already. Our decision did not mean that Europe wasn't important; however, in electronic ceramics Japan is where the big boys play and if you can't compete there you eventually won't be able to compete anywhere. But what approach should we take? Whatever we did there was an inherent danger that we could eventually lose our core technology in the process. Because of this, some of the management team wanted to finally bite the bullet and execute straight technology and know-how licenses with one or more companies that had already expressed an interest, including Murata, Kyocera, and Mitsubishi. They were willing to pay attractive royalties, any one would make an excellent second source, and each was capable of wiping us out long term once they had our technology. There were other large corporations who were not in components, but who had lots of cash and were looking to diversify. This was fine for further investment, might eventually lead to a viable second source, but offered little in the way of direct market penetration. I really believed we needed someone in the business and, based on the experiences with Nichicon-Sprague, Sanken, and MMCC, concluded that a joint venture in Japan, structured along the lines of Nichicon-Sprague, was the only alternative that made long term sense. The one thing I was not going to do was to just license away our birthright for quick return. Just as important, in any joint venture I was going to keep control of the U.S. market and try and find a way to get some form of direct interface with the Asian customers. For the moment, Europe was on the back burner due to limited resources. The real key would be to develop the right association with the right company under mutually beneficial terms—no easy task! After two months of heated internal debate, including several acrimonious board discussions, in the fall of 1989 "Jack" went off to Japan to reestablish contact with his old sources and others suggested by "Jimmie," put out some feelers, and see who might be the most appropriate partner.

As we approached the end of 1989, our first real production year, things were both looking up and getting more complicated. There was also some very good news: NEWCAPs and their relatives were better than even we had imagined, and there still was no real equivalent competition in the marketplace. The key to this success seemed to be the extraordinarily thorough and complete job that had been done from the beginning in characterizing and understanding the materials and processes that we used. Even "Jimmie" had commented that he had seen no better job done in Japan. "However, John, 'Marie' and I will show you how to do it even better!" They did.

"Jack" was gone for almost a month and, upon return, had identified close to a dozen possible partners, three of which seemed close to ideal. We finally settled on "Sumo Ceramics," a medium-sized company with revenues close to $100,000,000, primarily in specialty electronic ceramic components, and a niche position in high performance MLCs for the embryonic Japanese space effort. While their capacitor technology was adequate, because of the very low unit volumes involved, it was antiquated by today's standards. Still, they had an excellent reputation for the performance and quality of their products, knew Sprague Electric through its capacitor reputation, had evaluated some of NEWCO's MLCs, which they had apparently obtained through some of the restricted sampling we had done in Japan earlier in the year, and they seemed very interested in developing a relationship. While in no way a Murata, TDK, or Kyocera, they seemed much more suitable as a partner who could simultaneously provide an alternate source and market access in Japan, with much less chance of eventually overwhelming us. In addition, as far as we could tell, although they were well capitalized, they were not part of any industrial group or keiretsu. Several other companies were also chosen as potential back-ups and, in late November, "Jack" and I headed back to Japan to see if we could cut a deal.

Anyone who wants to succeed in Japan must have an iron constitution and the patience of Job. The number of trips we made, clubs we bonded in, and meetings we held in smoke-filled rooms became lost in a haze of frustration and fatigue. With a young company to run, we never could stay very long and they seemed bent on outlasting us. Several times we left meetings vowing to terminate the negotiations, although we never did. Finally, at the end of their second visit to the United States, Sumo-NEWCO was formed in late February of 1990. The ownership was split as 51% "Sumo" and 49% NEWCO. Our contribution was our powder and MLC technology, and they agreed to both capitalize the venture and manage it. In a rare change from similar ventures, "Sumo" agreed that we could have a resident manager in Japan whose primary responsibility was to establish direct contact with key customers, although the agreement specified that sales in Asia were to be handled through the "Sumo" sales organization. We were somewhat surprised to find that the Japanese customers actually liked this arrangement since it gave them more direct access to NEWCO who they recognized as the technology source. On the other hand, at least initially all sales in the U.S. were by NEWCO. However, as a quid quo pro for our joint venture resident, they assigned an engineer to work in our labs in North Adams. Because of his deep grounding in manufacturing and process engineering, "Charlie" turned out to be an excellent addition. As a matter of fact, the combination of the innovation skills of Ross, his small team, what was soon to become MRA Labs, and the commercialization and manufacturing competence of "Jimmie" and "Charlie" was a critical element in our future success.

Since neither "Sumo" or NEWCO had any real position in Europe, we agreed to work together on developing a joint presence there. Tight restrictions were negotiated to try and protect our technology from diffusion to other potential Japanese competitors. Since intercompany mobility is so restricted in Japan, I was much less worried about loss of our proprietary information than I would have been with such a venture in the United States. The venture started in excess factory space "Sumo" owned in one of the Kyoto suburbs, and, although it would be close to two years before there was appreciable Asian production, "Sumo" immediately began to seek Japanese opportunities for NEWCO MLCs and capabilities in specialty products. We also became the sales agent for "Sumo Ceramics" in the United States. As of this writing, while moving in a sometimes painfully slow and disciplined manner, there is every indication that the joint venture will be a rousing success. Pilot production has begun in Kyoto, two custom export programs are in development and design in North Adams, and NEWCO MLCs are orbiting the earth in several incredibly dense memory boards in a Japanese telecom satellite. As a member of the Sumo-NEWCO board of directors, I have even occasionally found myself back again at the Shell Club in Kyoto. For some reason it is more fun this time.

I need to make a brief comment concerning the ownership situation. As had been the case with Nichicon-Sprague, "Sumo" holds the majority position both in ownership and board representation. While this might seem a problem in the U.S., so far it has not been in our joint venture. No major decisions can be made without mutual consent of both "Sumo" and NEWCO, and the terms of the agreement have been consistently honored. The managements of the two companies also have considerable mutual respect, and the joint venture should be a success because both parent companies will benefit. Obviously, only time will tell.

MRA Labs was formed in the spring of 1990 and, over time, has become an important source of innovation and analytical and library services. During the latter part of the year, we also began to gradually shift to them as our emulsion source of $BaTiO_3$. In parallel, "Marie" began a detailed investigation of alternative sources of fine powder ceramics. Our technical team concentrated on further formulation development to tighten the TCC and increase volumetric efficiency through further decrease in the ceramic dielectric thickness. We also began to expand our formulation work into the Z5U temperature coefficient range. We felt this would permit us to more effectively go after the lower end of the solid tantalum capacitor market and to develop specialty high energy storage capacitors for switch mode power supplies. We were not looking for breakthroughs, although we knew they might come. Rather our goal was continuous incremental improvement in everything we did. Customer satisfaction gave every indication we were succeeding, and both the 1989 and 1990 financial results well exceeded plan. The real question for the future was how much effort and expense we could afford in more commodity-related MLCs.

Since we were both capacity and financially limited, we continued to emphasize the specialty side of the business and offer MLCs only in those applications where our superior characteristics allowed us to obtain higher margins. As time went on and "Sumo"-NEWCO became a more important factor, this would become an increasingly difficult strategy to follow. "Sumo Ceramics" saw a real possibility of selective penetration of the Asian market in commodity MLCs and was prepared to make a major related capital investment in the venture. In return, they began to press NEWCO to match at least part of their investment. This would lead to increased tensions in 1992 which have yet to be resolved.

1991 was another banner year and, like 1990, again exceeded our financial plan both in revenues and operating and net profit. In addition, because of the high margins of our product mix we were still able to stay well within our cash availability. To create a buffer against unpleasant financial surprises, we also negotiated a credit line with one of the local banks under reasonably attractive terms. I was very pleased at how well "Mike" handled these negotiations. It was evident that he was rapidly gaining the skills of a true CFO. Our reputation was also spreading and we began to receive feelers from several European firms concerning a variety of different forms of "strategic alliances." Europe was still untapped as a market to NEWCO, and we recognized that to serve the European Community meant that we would have to have local presence. On the other hand, as in Japan, I was unwilling to merely license our technology and again planned to pursue the joint venture route. "Sumo" agreed and hoped to also participate in such a venture. This became another item on our rapidly expanding agenda for 1992, or 1993 at the very latest.

Since the formation of NEWCO, we had held at least bimonthly meetings for all employees. This was relatively easy when the company was small and everyone was in the same location. They were held off-hours so as not to interfere with production, and this was first resented by some employees as an interference in their free time. Eventually, the practice was accepted by everyone as they realized how important such meetings were to involving them in the business. At these meetings we reviewed our performance versus customer commitments, quality record, the status of our TQC program which was now formalized throughout the organization, technology programs, financial performance and, in fact, almost every aspect of the business. These sessions were also used to openly discuss any other items participants wanted to bring up, as long as they related in some way to the affairs of NEWCO.

Early in February of 1992 just such a review was held at which we were able to report highly positive results in nearly every category. This was especially rewarding since NEWCO's growth was occurring at a time when the rest of the electronic components industry and general economy were still in the doldrums. The power of successfully pursuing a differentiated niche with a superior product capability was never more evident. However, there were also some significant concerns.

Although we had done an outstanding job with IBM, "Electro Systems," and our other specialty programs, because of limited resources we were in danger of accepting too many opportunities and of beginning to slip on customer commitments. In addition, there was increasing pressure from some of our good customers, "Jack" and his sales force, and from "Sumo Ceramics" to ramp up our MLC production capability, a move I was concerned would sap our limited human and financial capabilities. On the technical front, "Marie's" work had recently shown that hydrothermally synthesized fine-powder $BaTiO_3$, which was becoming commercially available from at least one Japanese source, might have the capability of producing results nearly competitive to the emulsion material. We still were well ahead in a number of our formulations and in our general processing capabilities. However, as I had originally feared might happen, one of our important technological advantages seemed to be eroding.

There were also several problems on the personnel front. Probably most important, Sid had indicated that he no longer wanted to work full time and that we should start to look for a replacement. There was not anyone in the organization or in MRA who could come even close to filling his shoes. In addition, because of the geographic isolation of the northern Berkshires, we had been unable to hire several talented and much needed BS engineers who wished to pursue their masters degree on a part time basis. This meant we really needed accelerate re-establishment of some of the cooperative programs that Sprague had once had with Williams College and with RPI.

However, the most serious potential crisis—which was known only to the top management team and not discussed at our review meeting—was the strong indication that "Electro Systems" might be considering the sale of our equity. This was due to a sharp downturn in their business fortunes during the latter part of 1991, which paralleled the continuing volatility of the overall computer industry. We had sensed that there was something wrong when they began to delay releases against their forecasted substrate requirements. In addition, while they continued to deny that there were any real problems, sources within the financial community supported our concern. Under our agreement, should they decide to sell, NEWCO had the right of first refusal and I wanted to be in a position to exercise this right should it become necessary. We still seemed far removed from being able to swing an initial public offering (IPO) and, because of this, it appeared that we were looking at finding another corporate partner. I was sure this was possible in Japan and "Sumo" seemed like a strong potential candidate, as did several other Asian companies. However, as far as ownership was concerned, I really wanted to go with a U.S. company if at all possible. Even with only a 25% equity position, companies such as IBM and other major U.S. equipment houses seemed just too large and I felt we would be lost in their huge corporate bureaucracies. This is exactly what happened to Mostek when they were acquired in their entirety by United Technologies. Therefore, again turning to my son Bill, we began to quietly create a list of viable, "Electro Systems-like" corporations that might be good partners. Hopefully ,exercising this list would not become necessary. Unfortunately, this hope was soon dashed when, in March of 1992, "Bill O'Neil" called to say that "E. S." had received an unsolicited offer from a large Japanese corporation to purchase their share of NEWCO's equity and that they intended on proceeding. While I questioned that any respectable Japanese company would really attempt such a purchase if NEWCO management were in opposition, it was clear that "Electro Systems" wanted out. Therefore, as far as I was concerned, we had 90 days to find an alternative. The remainder of our conversation is not worth recording!

There is a great deal of *deja vu* in this event and last night was a sleepless one. Mostek, Sprague, and now NEWCO? But life goes on and nothing is ever perfect or how one ideally wishes it should be. I called my son, told him the news, and asked his help in trying to identify and close with a new partner. From a business standpoint, such a deal is small potatoes for a Smith Barney. For a son trying to help his father, it is anything but. Next week I go on the road again, not to seek added business but to replace "Electro Systems." Failure is not a viable alternative.

EPILOGUE

Sprague Electric Redux

A great deal of the material in this book deals with Sprague Electric. Someday I hope to detail the complete story of its birth, growth, maturity, decline, and demise. You might wonder what happened to Sprague Electric and why it happened. There are some heroes, a few villains, but mostly the result was due to important management errors, some bad luck, and changes in the world economic environment and competitive situation.

On Friday, February 14, 1992, the Sprague Electric Company that I grew up with and worked for officially died, at least as I had known it. Already mortally wounded by restructurings, sale of assets, and exit from important growth business segments, the *coup de grace* was struck at a special meeting of stockholders when it was voted to sell the remaining core tantalum and U.S. thick film businesses to Vishay Intertechnology, Inc. While three small component operations remained at the time this book was being written, these were "on the block" as of mid-1992 and will likely have been sold by the time you read these words.

The future of the company became clearer in a June 26, 1992 press release. The STI Group (the then-current name for what had been Sprague Technologies, Inc., and before that Sprague Electric) announced it was considering a proposal from American Financial Corporation (AFC) to purchase AFC's wholly owned subsidiary, Great American Life Insurance Company, for $485,000,000. As part of the transaction, AFC was to increase its equity ownership in STI from 32.5% to 80%. Great American sells qualified tax-sheltered annuities to school teachers through payroll deduction programs.

The result of all this activity is that in the eleven short years since the 1981 acquisition of Sprague Electric by the Penn Central Corporation, the once dominant force in electronic components will become an insurance company.

That's right—*an insurance company!*

What happened?

Following World War II, Sprague was one of the most powerful component suppliers in the world. Flushed with success, it added semiconductors to its product portfolio, convinced that this was where the future lay. While this was to prove correct, such diversification was a strategic mistake for two important reasons. First, Sprague just did not have adequate resources to be a leader in both passive and active components. Secondly, the semiconductor thrust reoriented key management focus away from what would always be the core business, namely capacitors. If Sprague had continued to concentrate on passive components, it probably would have prevailed in MLCs, the key growth segment of capacitors.

Secondly, Sprague should have limited its 1960s employment in North Adams to 2000 employees or less. The 1970 strike that resulted from oversaturating the local labor market was an event from which Sprague never quite recovered.

The privatization that resulted from these two factors plus the mid-1970s recession was ultimately strongly negative. While ownership by General Cable/GK Technologies was stable (primarily due to a market that was also so), ownership by Penn Central was chaotic, at best. Fueled by the go-go years of the early 1980s, Sprague entered on an almost uncontrolled growth phase that would be disastrous when the market broke in 1985. Considered overly conservative "country boys" by the new owners, it was grow or go for Sprague management. So grow we did, only to start the restructuring phase in 1986, and eventually go in 1987 when Sprague Technologies, Inc. was formed as a spin-off to the Penn Central stockholders.

The new management team tried a "New Beginning," but it was really too late. Restructurings just couldn't keep up with falling margins and one after another growth businesses were shed due to inadequate returns. Finally, the once proud research and development organization was also gone and the final end was inevitable.

Could Sprague have been saved? Possibly, but only through a merger with one ore more other U.S. capacitor manufacturers. Possibly a consolidated "U.S. Capacitors, Inc." made up of Sprague, AVX, and Kemet might have saved the U.S. industry. Now this is obviously no longer an option.

What role did the Japanese play in all this, other than to become increasingly formidable competitors? It would be nice to be able to target them as the real culprits, and certainly they were no help. The capacitor dumping that occurred during the late 1960s and early 1970s hurt, but not fatally. More than blaming the Japanese, I blame the U.S.

government for lack of concern. Nichicon-Sprague was a net positive, while it lasted, although Sprague did not get out of the joint venture as much as it should. The Sanken debacle certainly hurt the semiconductor division, although it is not completely clear which side was most at fault.

Mitsubishi's MMCC was able to gain free access to all of Sprague's MLC capacitor technology without giving really anything back in return. Yet Sprague did this openly and knowledgeably, hoping somehow it would lead to a joint venture between the two companies. It didn't.

Perhaps the greatest harm done by the Japanese electronics industry was its impact on Sprague's U.S. customer base. The Japanese took away the entire U.S. consumer electronics industry, once a major Sprague customer. Japan's increasing penetration in the U.S. automotive industry is also effectively reducing this customer base as well. Only time will tell whether this will also happen in the all important computer segment.

Regardless of where the blame lies, the loss of Sprague is a tragedy for the U.S. electronics industry, the Sprague employees who have and will lose their jobs, the communities where Sprague once operated and does no longer, Sprague's customers and suppliers, and to me personally.

Notes and Sources

In addition to works previously cited in the text, information for this book was obtained from the following sources:

Chapter One

The sources for this chapter were private conversations with a number of people who, in one way or another, knew of or were involved in the Sprague Electric/Sanken Electric relationship.

Chapter Two

A number of different literature references were consulted, of which the following were most important:

Coursey, Philip R., *Electrical Condensers.* London: Pitman, 1927.

Brotherton, M., *Capacitors.* New York: Van Nostrand Company, 1946.

Britton, K. G., *Electronic Developments.* London: George Newnes Ltd., 1947.

Bendz, W. E. I., *Electronics for Industry.* New York: John Wiley & Sons, 1947.

Kloeffler, R. G., *Industrial Electronics and Control.* New York: John Wiley & Sons, 1949.

Moore, W. J., *Physical Chemistry.* New York: Prentice-Hall, Inc., 1955.

Overhage, Carl F. (editor), "The Age of Electronics" in *Lincoln Laboratory Decennial Lectures.* New York: McGraw-Hill Book Co., 1962.

Ryder, John D., *Electronic Fundamentals and Applications.* Englewood Cliffs, NJ: Prentice-Hall, Inc., 1964.

Bergamini, David, *Mathematics.* New York: Time, Inc., 1963.

Wilson, Mitchell, *Energy.* New York: Time, Inc., 1963.

Gilder, George, *Microcosm.* New York: Simon & Schuster, 1989.

Levinson, L. M., *Electronic Ceramics: Properties, Devices, and Applications.* New York: Marcel Dekker, Inc., 1988.

"50 Years of Achievement: A History," *Electronics,* April 17, 1980.

"The Transistor. . . The First 40 Years," *Solid State Technology,* December, 1987.

"How von Neumann Showed the Way," *American Heritage of Invention and Technology,* Fall, 1990.

"The 30th Anniversary of the I.C.: Thirty Who Made a Difference," *Electronic Buyers News,* September, 1988.

Chapter Three

A major source was my own knowledge from my years in the industry and input from my daughter, Cathy. Others consulted were Tom Prokopowicz and Galeb Maher of MRA Laboratories and Darnall Burks of Mitsubishi Materials. Levinson's *Electronic Ceramics* cited previously in Chapter Two was another useful source.

Chapter Four

The material on lifetime employment practices in Japan came from interviews conducted at the Electronics Industries Association of Japan (EIAJ) during my June, 1991 visit to Japan.

Chapter Five

My own experience and discussions with business associates were important inputs. Also valuable were "Tough Questions for U.S. as Companies Seek Ideas Overseas" in the January 1, 1992 New York *Times* and "Concurrent Engineering" in the July, 1991 issue of *IEEE Spectrum.*

Chapter Six

The materials distributed at the AMA/JMA conference mentioned at the beginning of this chapter were a useful resource. I received a great deal of help from John Winant in recounting the history of labor relations at Sprague Electric. Also useful were the articles "Asiapower" and "Reversing Sagging Precollege Skills in Mathematics and Science" in June, 1991, and December, 1990, respectively, issues of *IEEE Spectrum.* See also "Incentivize Me, Please" by D.W. Linden in the May 27, 1991 issue of *Forbes.*

Chapter Seven

Interviews were conducted with Jim Howard, manager of quality and reliability for Sprague Electric's solid tantalum operations as well as Larry Bromley, vice president of quality assurance at Aerovox, Inc. Additional insights were provided by the article "5 Deadly Sins of Quality Improvement" in the #3 1991 issue of *Ceramic Bulletin.* A particularly important text relative to the competitive situation in the global semiconductor industry is *Technological Competition in Global Industries: Marketing and Planning Strategies for American Industries* by David Methe' (Westport, CT: Quorum Books, 1991).

Chapter Eight

My experience at Sprague Electric was a major source. I found John Sheridan's series of articles on world class manufacturing in the July 2, 1990, August 6, 1990, and October 15, 1990 issues of *Industry Week* to be useful.

Chapter Nine

An interview with Neal Welch supplemented my business experiences in preparing this chapter.

Chapter Ten

My son Bill, who is a managing director at Smith Barney, was very helpful in trying to educate me about the cost of capital.

Chapter Eleven

"Japan, Inc.'s Flip Side" in the July 5, 1990 issue of the Boston *Globe* provided several insights for this chapter. Also useful were "Japan Says Sayonara to Womb-to-Tomb Management" by J. C. McClune in the November, 1990 issue of *Management Review* and "Japan Today" by J. Teresko in the September 2, 1991 issue of *Industry Week.* Insights into the possible results of "Europe 1992" were found in "Europower: Looking Ahead to 1992" in the June, 1990 issue of *IEEE Spectrum.*

Chapter Twelve

The previously-cited *Technological Competition in Global Industries* by Methe' was very useful in preparing this chapter.

Chapter Thirteen

While NEWCO is my strictly fictional creation, I did receive important input on its "financing" from my son, Bill, and Ed Getchell. Galeb Maher was also particularly helpful in developing the product strategy and in technically demonstrating that it was, in fact, feasible.

In writing this book, I also benefited from numerous personal contacts I have made over the years in the electronics and ceramics industries. While these people do not necessarily agree with the opinions I express in this book, they all provided insights and observations that were a valuable stimulus to my own thinking. I wish to acknowledge and thank the following people:

ALCOA, Alcoa Center, PA
- Dr. Gordon R. Love, technical director–ceramics

Allegro Microsystems, Worcester, MA
- Dr. Kenneth Manchester, consultant
- Richard J. Morrison, president

AVX Corporation, New York, NY
- Peter R. Loconto, president

Electronic Industry Association of Japan, Tokyo, Japan
- Shoichi Miki, manager, electronic components department
- Shizuo Ogura, director, electronic components department
- Iwao Ojima, president

Konica Corporation, Tokyo, Japan
- Dr. Sei-ichi Denda, managing director

Matsushita Electronic Components Co. Ltd., Tokyo, Japan
- Robert Yoshida, director, Tokyo business management department

Mitsubishi Materials Corp. Ceramics Division, Chichibu, Japan
- Darnall P. Burks, technical advisor
- Mikiya Ono, managing director and general manager
- Shungo Saitoh, planning department

MRA Laboratories, Inc., North Adams, MA
- Dr. Galeb H. Maher, president and chief executive officer
- Thomas I. Prokopowicz

Murata Erie North America, Inc. (MENA), Smyrna, GA
- Jack Driscoll, senior vice president, marketing and sales

Murata Manufacturing Co., Ltd., Kyoto, Japan
- Yasutaka Murata, president
- Dr. Michihiro Murata, general manager, Yokohama research and development center
- Mitsuru Sawa, chief, overseas support
- Motohisa Yumoto, assistant to the senior executive director, general administration department

Nichicon Capacitor, Ltd., Kyoto, Japan
- Ike Takeda

Nissan Motor Company, Ltd., Yokosuka, Japan
- Dr. Ryoichi Nakagawa, research consultant

Sharp Corporation, Osaka, Japan
- Dr. Tadashi Sasaki, senior advisor and president, Executive Networks Japan, Inc., Tokyo, Japan
- Akira Fujii, Executive Networks Japan, Inc.

Sprague Technologies, Inc./Sprague Electric Company, Stamford, CT
- Peter W, Maden, vice president, solid tantalum division

Techno-Venture Company, Ltd., Tokyo, Japan
- Dr. Yaichi Ayukawa, president
- Haruyoshi Muratsubaki, investment advisor
- Fumiaki Sudo, advisor to the president

Index